D0926636

THE CREATION OF

BRIDGES

THE CREATION OF BRIDGES

From vision to reality - the ultimate challenge of architecture, design and distance

DAVID BENNETT

AURUM

A QUINTET BOOK

First published in Great Britain
1999 by Aurum Press Ltd
25 Bedford Avenue, London WC1B 3AT

A catalogue record for this book is available from the British Library.

ISBN 1-85410-651-1

1 3 5 7 9 10 8 6 4 2
1999 2001 2003 2002 2000

This book was designed and produced by
Quintet Publishing Limited
6 Blundell Street
London N7 9BH

Creative Director: Richard Dewing
Designer: James Lawrence
Senior Editor: Sally Green

Typeset in Great Britain by
Central Southern Typesetters, Eastbourne
Manufactured in Hong Kong by
Regent Publishing Services Ltd
Printed in China by
Leefung-Asco Printers Ltd

Picture Credits

Key Numbers refer to page numbers. *b* = bottom, *l* = left, *m* = middle, *r* = right, *t* = top.

Margaret Amman-Durer 201*t*, 202*b*; **Boilly Photo, Prince Edward Island, Canada** 222*tl*; **British Cement Association** 55*mr*, **Caltrans** 47*t*, 136, 152; **Julie V Clark** 41*tr*, 63, 64*bl*, 66, 67, 81*r*, 87, 154, 216*b*; **Construction News** 173; **Robert Cortright** 15*ml*, 14*m*, 17, 20*b*, 21, 25*m*, 34*tr*, 41*ml*, 43*br*, 53, 55*t*, 57*bl*, 71*t*, 101, 104, 105, 134, 157, 169, 184, 187, 189*t*, 192*t*; **Jenny Crossley** 22*b*, 23, 109*m*, 143, 225*t*; **Eric Delony** 27*b*, 29, 35, 41*b*, 43*br*, 44*t*, 64*bl*, 71*b*, 77*b*, 137, 167, 201*b*, 205; **ETH Zurich** 73; **FBM Studio** 55; **Hugh Ferguson** 141*t*, 171; **Flint & Neil Partnership** 52*tl*, 65*r*, 140, 153, 170, 172; **Dave Freider** 39, 128, 202*t*, 207; **Freyssinet** 54, 57*br*, 70*b*, 74*m*, 75*t* & *r*, 79*m*, 121, 122, 208, 209, 210, 211, 212; **Freyssinet/photographer Francis Vigouroux** 28, 52*b*, 58*b*, 59, 64*b*, 79*b*, 80, 93, 94, 95, 96, 97, 127, 145*t*, 146, 148, 149, 150, 151; **Jolyon Gill** 31*t*, 186; **Honshu-Shikoku Bridge Authority** 49*tr*, 81*l*; **Institution of Civil Engineers** 9, 11, 18, 27*t*, 32, 33, 99, 108, 109*t*, 110, 111, 112, 113, 118, 119, 120, 138, 139*t*, 158, 159, 161, 162, 163, 165, 166, 168, 177, 178, 179, 180, 181*b*, 182, 183, 185, 188, 189*b*, 190*t*, 191*t*, 192*t*, 193; **Jean Muller International** 57, 58*tl* & *r*, 98, 142, 144, 175, 219, 220, 221, 222, 223, 224, 225*b*, 226*t*, 227; **Fritz Leonhardt** 30*m* & *b*, 49*tl*, 51, 52*mr*, 56*tr*, 57*ml*, 64*tl*, 65*tl*, 190*b*, 191*b*, 213*b*, 214*b*, 215, 216*t*, 217*t*, 218; Life File 197; **MTA Bridges & Tunnels Special Archive** 50, 90*bl*, 91*r*; **J-L Michotey** 16, 22*t*, 44*b*, 72*t*, 77*tr*, 103, 109*b*, 115, 181; **New York Port Authority** 47*br*, 129, 203, 204; **Ove Arup & Partners** 56*b*; **Jorg Schlaich** 10; **Grant Smith** 19, 41*tl*, 123, 130, 131, 132, 133, 226*b*; **Alain Spielmann** 74*b*, 106, 107, 116; **Steinman Consulting Engineers** 36, 37, 38, 41, 42*tl* & *bl*, 43*tr*, 45, 46*bl*, 47*bl*, 48, 49*m*, *b*, 76, 78*tl*, *tr*, 83, 85, 88*b* & *r*, 89*r*, 90*tl*, *br*, 135; **Denton Taylor** 174, 193, 195, 194, 197, 200; **Michel Virlogeux** 13, 79*tr*, 114; **Hans Wittfoht** 8, 12*t*, 14*t*, 42*bl*, 64*mr*, 100*b*, 214*t*.

Contents

Foreword

Bridges have been and still are the umbilical cords of humankind's progress over the centuries. From the simple log and the crude rope suspension bridge to the monumental concrete and steel spans of today, bridges have dramatically unfurled the history of humanity's conquest of nature's barriers—a river, a chasm, an estuary, a valley, and even the sea. Bridges just a few feet wide to many miles long have the same common goal of serving the need for better access, better transportation links and trade between local communities and international boundaries.

In the past century there has been more bridge building, more discovery of new technology and advances in bridge-engineering science than in all the preceding centuries put together. The year 1998 has been a bumper year for bridge-building activity, on a scale so vast and so financially huge that it is difficult to comprehend. In Japan alone the current bridge-building program is the equivalent of building all of New York's bridges in one go. Just imagine the Brooklyn, the George Washington, the Verazzano Narrows, the Bronx Whitestone, Throgs Neck, Manhattan, Williamsburg, Queensboro, the Goethals, Hellgate, Bayonne, and Outerbridge all built one after the other and within ten years. It beggars belief, but it is happening.

Why this urgency? The answer is space for housing for a rapidly increasing population, and for the future needs for commerce and transportation, between the Honshu mainland and the island of Shikoku. The Japanese government are building an infrastructure that must serve the whole community and stimulate economic growth in Japan for the next hundred years.

In charting the history of bridge building through the centuries, recalling some of the greatest bridges ever built and the horrific tales of the worst bridge tragedies, one thing is dominant and common among them all. No matter how big or how small they may be, every bridge project must start with a vision of its creation, followed by the endeavor to make that creation a reality. Bridges were not built as monuments for pleasure or grandeur but as an economic necessity in the service of a community or nation. They blazed a trail over inhospitable lands, over rapid-flowing streams and deep gorges.

This book explains by simple diagrams and everyday experiences the fundamental principles behind the design and construction of a bridge, from post-and-lintel and arch-and-truss bridges to box girders, cable-stay, and suspension bridges. The many illustrations and examples of the different types of bridges will help you to recognize them during your travels. The stunning locations and elegance of many of the bridges shown may also inspire you to make a special journey or detour while on vacation or business to see some of these man-made wonders.

You will see many unfamiliar terms as you read on. To help, I have provided a glossary at the back, which explains most of the terms. Others should become apparent from context.

I have one or two regrets in researching my material: I could not find space to put everything in. I have not been able to include material about footbridges, swing bridges, lifting bridges, and transporter bridges, nor say much about the current controversy in the bridge world over who should conceptually design a bridge—an architect or an engineer.

I wish to acknowledge the help I received from Freyssinet International with copies of their publications and the many images of bridges; Jean Muller International (JMI) for information and the many images on JMI bridges; Michael Chrimes at the ICE Library in London; Eric DeLony of HAER; George Gesner of Steinman Consulting Engineers for archive material; Alain Spielmann, Michel Virlogeux, Jean Muller, Fritz Leonhardt, and Angus Low for their views on bridges.

The inspiration for this book has been sourced from David Steinman's wonderful book *Bridges and Their Builders*, Joseph Gies's *Bridges and Men*, Hans Wittfoht's highly informative *Bridge Building* and Mario Salvadori's thought-provoking *Building—from caves to skyscrapers*.

David Bennett, October 1998

1

The early history of bridge building —the age of timber and stone

The bridge has been a feature of human progress and evolution ever since man the hunter-gatherer became curious about the world beyond the horizon and the fertile land, the animals and fruit flourishing on trees on the other side of a river or gorge. Early human beings had to devise ways to cross a stream and a deep gorge to survive.

A boulder or two dropped into a shallow stream works well as a stepping stone, as many of us have discovered—but, for deeper-flowing streams, a tree dropped between banks is a more successful solution. So the primitive idea of a simple beam bridge was born.

Although it is certain that early humans lived in groups and passed on such primitive technology, it is likely that the skill of making rudimentary bridges was discovered and rediscovered by succeeding generations many times over, until it became established as a skill. For those groups who lived in the forests, surviving on fruit, grubbing up roots, and snaring or spearing animals, ease of travel through the forest canopy was vital to locate new food sources and shelters.

ABOVE: A rope suspension bridge in Asia.

OPPOSITE: Primitive log bridge, Afghanistan.

ABOVE: An early post and lintel stone "clapper" bridge, Dartmoor, England.

Today, in the forests of Peru and the foothills of the Himalayas, crude rope bridges span deep gorges and fast-flowing streams to maintain pathways from village to village for hill tribes. Such primitive rope bridges evolved from the vine and creeper that early humans would have used to swing through the forest and to cross a stream. We have only to see a gibbon on a wildlife program or a chimpanzee at play in a zoo to get the idea of how it would work. Here is the second basic idea of a bridge: the suspension bridge.

For thousands of years during the Paleolithic period (2,000,000–10,000 BC), we know that our ancestors lived as nomads and wanderers, hunting and gathering food. Slowly it dawned on early man that following herds of deer or buffalo—or just foraging for plant food haphazardly—could be better managed if the animals were kept in herds nearby and plants were grown and harvested in fields. Regular routes between settlements became necessary to barter for grain and food stocks—and invariably to use when stealing from each other! If there was a shorter way to travel between two places, human enterprise would find a way to bridge a river or to cut a clearing through a forest.

In this period the simple log bridge had to serve many purposes. It needed to be broad and strong enough to take cattle; it needed to be a level and solid platform to transport food and other materials; and it needed to be movable so that it could be withdrawn to prevent enemies from using it.

Narrow tree trunk bridges were inadequate and were replaced by double-log beams spaced wider apart on which short lengths of logs were placed and tied down to create a pathway. The pathways were planed by sharp scraping tools or axes in the Bronze Age and any gaps between them plugged with branches and earth to create a level platform.

For crossings over wide rivers, support piers were formed from piles of rocks in the stream. Sometimes stakes were driven into the river bed to form a circle and then filled with stones, creating a crude cofferdam—a watertight, dry enclosure. Around 3500 BC early Bronze Age "lake dwellers" lived in timber houses built out over the lakes, in the area that is now Switzerland. To ensure the house did not sink, they evolved ways to drive timber piles into the lake bed. From the discovery of this came the timber-pile bridge and the trestle bridge.

For tribal groups living in the more northern glacial regions, with a plentiful supply of stones of all shapes and lengths and not as many large trees, the stone slab was preferred as a beam to bridge rivers and streams. Crude stone bridges that survive today in Dartmoor, England, and built around the fifteenth century, are reminiscent of prehistoric stone-bridge construction of the Bronze Age.

So primitive bridges were essentially post-and-lintel structures, made either from timber or stone or from a combination of both. Sometime later the simple rope-and-bamboo suspension bridge was devised, which developed into the rope suspension bridges that are in regular use today in the mountain reaches of China, Peru, Colombia, India, and Nepal.

It took human ingenuity till about 4000 BC to discover the secrets of arch construction. In the Tigris–Euphrates valley the Sumerians began building with adobe—a sun-dried mud brick—for their palaces, temples, ziggurats, and city defenses. Stone was not plentiful in this region and had to be imported from Persia, so was used sparingly.

BELOW: An engraving of an early Indian pontoon bridge.

The brick module dictated the construction principles they employed, to scale any height and to bridge any span. And, through trial and error, it was the arch and the barrel vault that were devised to build their monuments and grand architecture at the peak of their civilization. The ruins of the magnificent barrel-vaulted brick roof at Ptsephon and the Ishtar Gate at Babylon are a reminder of Mesopotamian skill and craftsmanship.

Although most Egyptian building was dominated by post-and-lintel stone construction, the corbeled arch had been discovered and was used frequently in constructing passageways, relieving arches, and escape tunnels within massive pyramid structures. At Dindereh today three arch ruins still stand that date from 3600 BC. By the end of the Third Dynasty, around 2475 BC, the Egyptians had mastered the true arch and it was in common use in construction.

Without doubt the arch is one of the greatest discoveries. The arch principle was the vital element in all building and bridge technology over later centuries. Its dynamic and expressive form gave rise to some of the greatest bridge structures ever built.

Earliest records of bridges

The earliest written record of a bridge appears to be one built across the Euphrates around 600 BC as described by Herodotus, the fifth-century Greek historian. The bridge linked the palaces of ancient Babylon on either side of the river. It had 100 stone piers, which supported wooden beams of cedar, cypress, and palm to form a carriageway 35 feet wide and 600 feet long. Herodotus mentions that the floor of the bridge would be removed every night as a precaution against invaders.

In China, it seems that bridge building evolved at a faster pace than the ancient civilization of Sumeria and Egypt. Records exist from the time of Emperor Yoa in 2300 BC on the traditions of bridge building.

Early Chinese bridges included pontoons or floating bridges and probably looked like the primitive pontoon bridges built in China today. Boats called sampans, about 30 feet long, were anchored side by side in the direction of the current and then bridged by a walkway.

The other bridge forms were the simple post-and-lintel beam, the cantilever beam, and rope suspension cradles. Timber beam bridges, probably like those of Europe, were often supported on rows of timber piles of soft fir wood called "foochow poles," so named because they were grown in Foochow. A team of builders would hammer the poles into the river bed using a cylindrical stone fitted with bamboo handles. A short crosspiece was fixed between pairs of poles to form the supports that carry timber boards, which were then covered in clay to form the pathway over the river.

ABOVE: Cantilever timber bridge in Tibet.

For longer spans and for bridges that must maintain a wide navigation channel, the cantilever beam or rope suspension bridge was the usual choice. The cantilever bridge, being rigid in construction, was preferred in the towns and cities and for river crossing along important trading routes.

BELOW: Illustration of a covered cantilever truss bridge in medieval China, which became the focal point for trade in a town.

To build a cantilever bridge, first a wooden caisson—a structure for keeping water out while deep foundations are being excavated—was formed on each bank and filled with stone, rubble and clay. Then timbers were driven into the front of the caisson and embedded deep in the fill to form a platform of cantilevers that spring from each bank. The two arms of the cantilever were then bridged over the central gap by simple beams.

As stone cutting and masonry construction was better understood, the wooden caisson was replaced by a stone abutment, which was often built into a house or gateway structure leading onto the bridge span. If a central pier was needed, it was often adorned with a pagoda or canopy structure to serve as a meeting place and an open market for buying and selling goods. The town bridge over a wide river became the town square in Chinese society and the center for commerce and trade.

In later centuries Chinese bridge building was dominated by the arch, which they copied and adapted from the Middle East as they traveled the silk routes that opened during the Han Dynasty around AD 100.

Through Herodotus we learn about the Persian ruler Xerxes (*c.* 519–465 BC) and the vast pontoon bridge he had built, consisting of two parallel rows of 360 boats, tied to one another and to the bank and anchored to the bed of the Hellespont, which is the Dardanelles today. Xerxes wanted to get his army of 2 million fighting men and horses to the other side to meet the Greeks at Thermopylae. It took seven days and seven nights to get the army over to the other side.

Xerxes' massive army was defeated at the Battle of Thermopylae in 480 BC, the remnants of which retreated back over the pontoon bridge to fight another day. The Persians were great bridge builders and built many arch, cantilever, and beam bridges. There is a bridge still standing at Dizful in Khuzistan, Iran, over the river Diz, which could date anywhere from 350 BC to AD 400. The bridge consists of 20 voussoir arches (formed from wedge-shaped stones) which are slightly pointed, hence the Gothic affinity, and has a total length of 1,250 feet. Above the level of the arch springing are small semicircular spandrel arches, which give the entire bridge an Islamic look, hence the uncertainty of its Persian origins.

The Greeks did not do much bridge building in their illustrious history, being a seafaring nation that lived on self-contained islands and in feudal groups scattered across the Mediterranean. They used exclusively post-and-lintel construction in

BELOW: Kintai bridge, Japan, built in 1673, is typical of many early Chinese and Japanese arch bridges.

evolving a classical order in their architecture, and built some of the most breathtaking temples, monuments, and cities the world has ever seen, such as the Parthenon, the Temple of Zeus, the city of Ephesus, Miletus, and Delphi, to name but a few. They were quite capable of building arches like their forebears the Etruscans when they needed to. There are examples of Greek voussoir arch construction that compare to the Beehive Tomb at Mycenea, like the ruins of an arch bridge with a 27-foot span at Pergammon in Turkey.

TOP: Partially destroyed stone arch bridge in Iran (Persia), built by the Sassanids (c400 AD).

ABOVE: Roman Aqueduct of Segovia in Spain, built during the end of the first century AD.

The Romans

The Romans on the other hand were the masters of practical building skills. They were a nation of builders, who took arch construction to a science and high art form during their domination of Mediterranean Europe. Their influence on bridge-building technology and architecture has been profound.

The Romans conquered the world as it was known then, built roadways, canals, and cities that linked Europe to Asia and North Africa, and produced the first true bridge engineers in the history of civilization. Romans understood that the establishment and maintenance of their empire depended on efficient and permanent communications. Building roads and bridges was therefore a high priority.

The Romans also realized, as did the Chinese in later centuries, that timber structures, particularly those embedded in water, had a short life and were prone to decay, insect infestation, and fire hazards. Prestigious buildings and important bridge structures were therefore built in stone. But the Romans had also learned to preserve their timber structures by soaking timber in oil and resin as a protection against dry rot, and coating them with alum for fireproofing. They learned that hardwood was more durable than softwood, and that oak was best for substructure work in the ground, alder for piles in water; while fir, cypress, and cedar were best for the superstructure above ground.

They understood the different quality of stone that they quarried. Tufa, a yellow volcanic stone, was good in compression but had to be protected from weathering by a stucco—a lime wash. Travertine was harder and more durable and could be left exposed, but was not very fire-resistant. The most durable materials, such as marble, had to be imported from distant regions of Greece and even as far as Egypt and Asia Minor (Turkey). The Romans' big breakthrough in material science was the discovery of lime mortar and pozzolanic cement, which was based on the volcanic clay that was found in the village of Pozzuoli. They

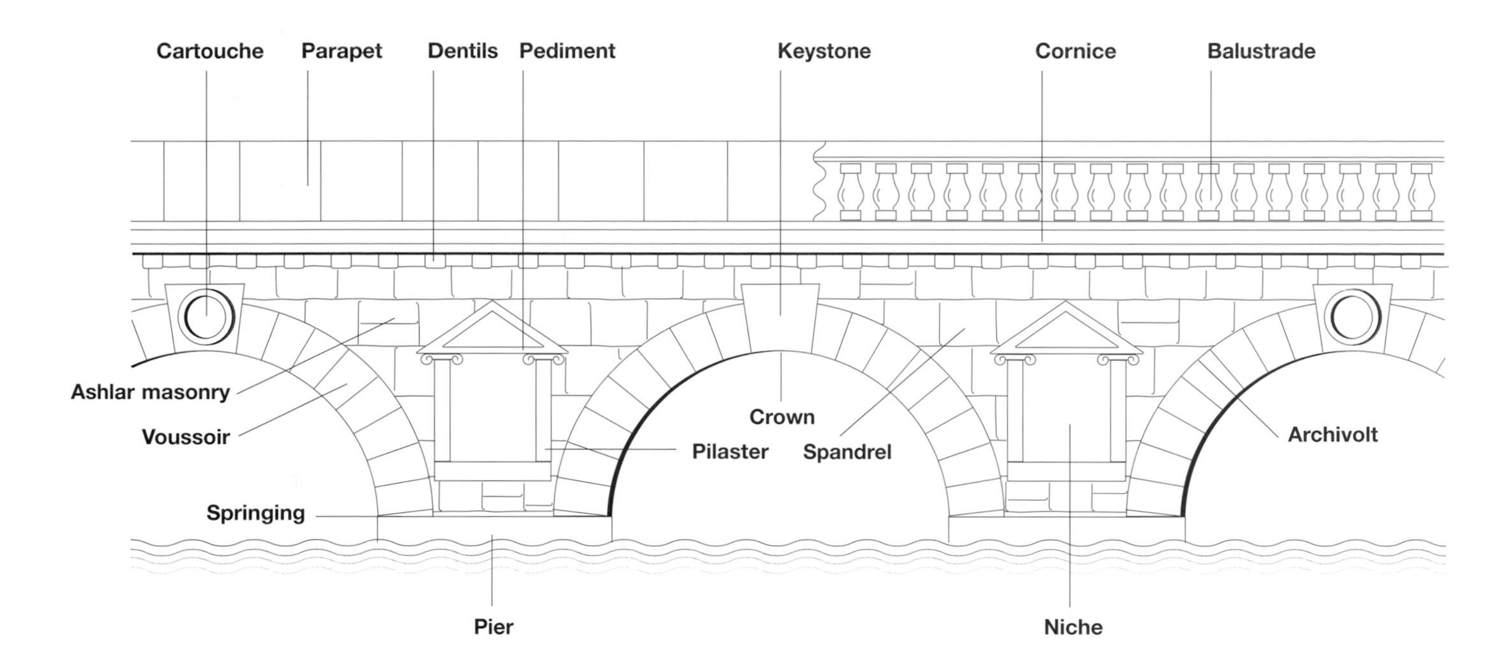

ABOVE: Architectural terms used in Roman and later masonry arch construction.

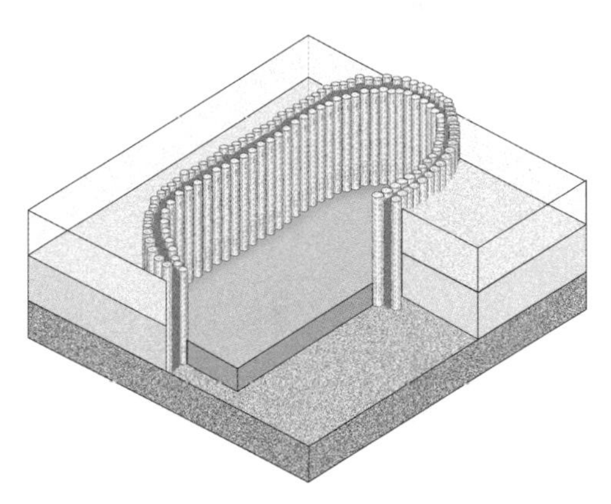

ABOVE: Roman coffer dam formed by two rows of timber piles, which are in-filled with clay to create a watertight enclosure.

ABOVE LEFT: Pons Fabricus in Rome, built in 62 BC. The modern name of the bridge is Ponte Quattro Capi, named after the branch of the Tiber that it spans.

used it as mortar for laying bricks or stones and often mixed it with burnt lime and stones to create a waterproof concrete.

The Romans realized that voussoir arches could span farther than any unsupported stone beam, and would be more durable and robust than any other structure. They ought to have known because the early Roman leaders and settlers were Etruscans. Semicircular arches were always built by the Romans, with the thrust from the arch going directly down onto the support pier. It meant that piers had to be large. If they were built wide enough at about-one third of the arch span, then any two piers could support an arch without shoring or propping from the sides.

In this way it was possible to build a bridge from shore to shore, a span at a time, without having to form the entire substructure across the river before starting the arches. They developed a method of constructing the foundation on the river bed within a cofferdam or watertight, dry enclosure, formed by a double ring of timber piles and clay packed into the gap between them to act as the water seal. The water inside the cofferdam was then pumped out and the foundation substructure was built within it. The massive piers often restricted the width of the river channel, increasing the speed of flow past the piers and increasing the scour action. To counter this the piers were built with cutwaters, which were pointed to cleave the water so it would not scour the foundations.

The stone arch was built on a wooden framework built out from the piers and known as centering The top surface was shaped to the exact semicircular profile of the arch. Parallel arches of stones were placed side by side to create the full width of the roadway. The semicircular arch meant that all stones were cut identically and that no mortar was needed to bind them together once the keystone was locked in position. The compression forces in the arch ensured complete stability of the span.

Of course the Romans did build many timber bridges but they have not stood the test of time, and today all that remains of their achievement after 2,000 years is a handful of stone bridges in Rome, and a few scattered examples in France, Spain, North Africa, Turkey, and other former Roman colonies. But what still stand today, whether bridges or aqueducts, rank among the most inspiring and noble bridge structures ever built, considering the limitation of their technology.

BELOW: Pont du Gard, Nîmes, France (c20 AD).

In Rome there are six bridges to be seen, the most celebrated of which is the highly decorated Ponte Sant'Angelo (136 BC) although many would regard the Pons Augustus in Rimini (AD 20) as the finer example, because of its classic proportions. Of the great aqueducts there is the Segovia Aqueduct (AD 1) in Spain, with its two tiers of 109 arches carrying the Rio Frio the last 2,500 feet into the town, and the most famous one of them all, the Pont du Gard at Nîmes, built in 19 BC. Its scale and monumental presence continue to impress.

ABOVE: The remains of Pont d'Avignon over the Rhône at Avignon, in France (1188).

The Dark Ages and the Brothers of the Bridge

When the Roman Empire collapsed it seemed that the light of progress around the world went out for a long while. The Huns, the Visigoths, Saxons, Mongols, and Danes did not do much building in their raids across Europe and Asia, plunder and destruction being higher on their agenda. It was left to the spread of Christianity and the strength of the church to start the next boom in road building and bridge building around AD 1000. It was the church that had preserved and developed both spiritual understanding and the practical knowledge of building during this period. This building naturally included the building of bridges.

In northern Italy there lived a group of friars of the Altopascio order near Lucca, in a large dwelling called the Hospice of St James. The friars were skilled at carpentry and masonry, having built their own priory. The surrounding countryside was wild and dangerous, and the refuge they built was a popular resting place for pilgrims and travelers using the ancient road from Tuscany to Rome.

In 1244 Emperor Frederick II required that the hospice build a proper bridge across the White Arno for pilgrims and travelers. Obviously the Hospice of St James was profiting well from passing trade for the Emperor to issue such a decree. With their skills they set up a cooperative to build the bridge.

After they had completed the bridge over the White Arno their fame spread through Italy and France. It sparked off an interest in bridge building among other ecclesiastical orders. In France a group of Benedictine monks established the religious order of the Frères Pontiffes (Brothers of the Bridge) to build a bridge over the Durance.

ABOVE: Engraving of "Old London Bridge" in the seventeenth century.

And so the Brothers of the Bridge order became established among Benedictine monks and spread from France to England by the thirteenth century. The purpose of the order, apart from its spiritual duties, was to give aid to travelers and pilgrims, to build bridges along pilgrim routes or to establish boats for their use, and to receive them in hospices built for them on the bank. The Brothers of the Bridge were great teachers, who strove to emulate and continue the magnificent work of the Roman bridge builders.

The most famous and legendary bridge of this period was built by the Order of Saint Jacques du Haut Pas, whose great hospice once stood on the banks of the Seine in Paris on the site of the present church with that name. They built the Pont Esprit over the Rhône, but their masterpiece was the neighboring bridge at Avignon. It was truly a magnificent and record-breaking achievement for its time. Its beauty has inspired writers, poets, and musicians over the centuries. What, then, was so special about the bridge? For a start the arch was not semicircular but elliptical in shape and therefore could span farther than a semicircular arch. It was more stable and could be made more slender over the crown. The result of all this was that the piers could be made narrower and the arch taller, thereby carrying the roadway higher out of reach of potential flooding and better for navigation. Small relieving arches were formed above the piers and in the spandrels to accommodate spring flood waters.

To have designed such a bridge you would have to be a mathematical wizard or have received divine inspiration from somewhere. The Pont d'Avignon comes with a legend about its shepherd-boy designer, Bénézet, who is said to have had a vision from God in 1178 commanding him to build it. When the Bishop of Avignon demanded proof of the boy's claim to divine intervention, the boy allegedly miraculously picked up an enormous boulder and carried it to the place where the bridge was to be built.

ABOVE: Ponte Vecchio in Florence, Italy (c1345).

The sanguine explanation is that Pont d'Avignon was masterminded by Brother Benoit, who supervised the brothers in the building of the Saint Esprit bridge and many other bridges. For its overall length of 1,300 feet and 20 spans, the bridge was remarkably slender and took only ten years to build, which is fast by medieval standards. The width of the bridge was also interesting. At its widest point it was 16 feet, but where the chapel was built over the second pier it narrows down to just 6 feet 6 inches. Such bottlenecks were designed by medieval bridge builders in order to defend the bridge more easily or perhaps to ensure a pilgrim toll was paid.

Sadly all that remains today at Avignon are just four out of the 20 spans of the bridge and the chapel where the supposed creator of the bridge was interred and later canonized as Saint Bénézet.

While Pont d'Avignon was being built in France, another monk of the Benedictine order in England, Peter of Colechurch, was planning the building of the first masonry bridge over the Thames. A campaign for funds was launched with enthusiasm. Not only did the rich townspeople, the merchants, and money lenders make generous donations, but the common people of London all gave freely. Until the sixteenth century a list of donors could be seen hanging in the chapel on the bridge. The structure that was built in 1206 was Old London Bridge and ranks after Pont d'Avignon in fame. It was such a popular bridge that

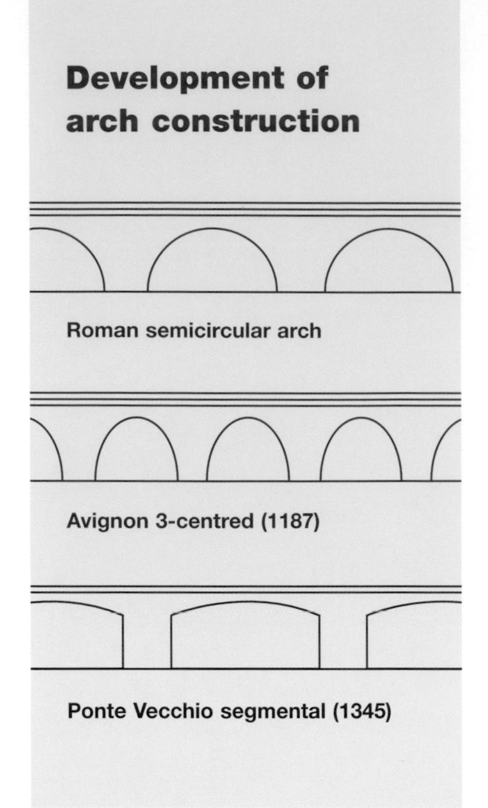

buildings and warehouses were soon erected on it. It became so fashionable a location that the young noblemen of Queen Elizabeth's household resided in a curious four-story timber building imported piece by piece from Holland, called the "Nonesuch House."

Towns continued to sponsor and promote the building of stronger and better bridges and roads. They did not always get the Brothers of the Bridge to build them, because they were often committed to other projects for many years. Instead, guilds of master masons and carpenters were formed and spread across Europe offering their services. Even government officials were united in this community enterprise and began to grasp the initiative and drive for better road and bridge networks across the country.

Soon the vestiges of the Dark Ages and feudalism were transformed to the Age of Enlightenment and the Renaissance. The Ponte Vecchio in Florence built toward the end of this period marks the turning point of the Dark Ages. It was a covered bridge erected in 1345, lined with jewelers' shops and galleries, with an upper passageway added later, that was a link between the royal palaces and those of government—the Uffizi and Pitti. The piers, which are 20 feet thick, support the overhanging building as well as the bridge spans. The most innovative features of the bridge are the arch spans, which are extremely shallow compared with any previous arches ever built or indeed many contemporary European bridges. It was built as a segmental arch, which is unusual for bridge builders of that period because they could not possibly determine the thrust from the arches mathematically, with the knowledge they had. How they did it is a mystery, as is the segmental arch of Pont d'Avignon. The architect of this radical design was Taddeo Gaddi, who had studied under the great painter Giotto, and was regarded as one of the great names of the Italian Renaissance that followed.

RIGHT: Example of a medieval fortified bridge. Monmow Bridge, Monmouth, Wales (1272).

The Renaissance

Not since the days of Homer, Aristotle, and Archimedes in Hellenistic times have such great feats of discovery in science and mathematics, and such works of art and architecture, been achieved as during the Renaissance. Modern science was born in this period through the inquiring genius of Copernicus, Leonardo da Vinci, Francis Bacon, and Galileo and in art and architecture through Michelangelo, Brunelleschi, and Palladio. During the Renaissance there was a continual search for the truth, explanations of natural phenomena, greater self-awareness, and rigorous analysis of Greek and Roman culture. As far as bridge building was concerned, particularly in Italy, it was regarded as a high art form.

As much emphasis was placed upon its decorative order and pleasing proportions as on the stability and permanence of its construction. Bridge design was architect-driven for the first time with da Vinci, Palladio, Brunelleschi, and even Michelangelo all experimenting with the possibility of new bridge forms. The most significant contribution of the Renaissance was the invention of the truss system, developed by Palladio from the simple king-post and queen-post roof truss, and the founding of the science of structural analysis with the first book ever written on the subject by Galileo Galilei entitled *Dialoghi delle Nuove Scienze* (*Dialogues on the New Science*) published in 1638.

ABOVE: Ponte degli Alpini built in 1568, one of the few Palladio timber truss bridges still standing today at Bessano, Italy.

Palladio did not build many bridges in his lifetime; many of his truss bridge ideas were considered too daring and radical and his work lay forgotten until the eighteenth century. His great treatise published in 1520—*Four Books Of Architecture*, in which he applied four different truss systems for building bridges—was destined to influence bridge builders in future years when the truss replaced the Roman arch as a principal form of construction.

Other groups of bridge builders during the Renaissance were clever material technologists who were preoccupied with the art of bridge construction and how they could build with less labor and materials. It was a time of inflation when the price of building materials and labor was escalating. The most famous bridge builders in this era were Ammannati, Da Ponte, and Du Cerceau.

Which bridge of the Renaissance is the most beautiful? Santa Trinità in Florence? The Rialto in Venice? The Pont Neuf in Paris? Arguably the most

famous and celebrated bridge of the Renaissance was the Rialto bridge designed by Antonio Da Ponte in Venice . "The best building raised in the time of the Grotesque Renaissance, very noble in its simplicity, in its proportions and its masonry." So said John Ruskin about the Rialto. Its designer was 75 years old when he won the contract to build the Rialto, and was 79 when it was finished. It is a single segmental arch span of 87 feet 7 inches, which rises 25 feet 11 inches at the crown. The bridge is 75 feet 3 inches wide, with a central roadway, shops on both sides, and two small paths on the outside, next to the parapets. Two sets of arches, six each of the large central arch, support the roof and enclose the 24 shops within it. It took three and a half years to build, and all the city's stonemasons were kept busy for two years.

It was also of a very novel construction, as you can see in Chapter 4, where I've gone into some detail in the section that looks at my pick of what I consider to be some of the best bridges in the world.

Equally innovative and skillful bridge construction was being progressed across Europe. In the state of Bohemia across the Moldau at Prague was built the longest bridge over water, the Karlsbrücke, in 1503, which was the most monumental and imperial bridge of the Renaissance. It took a century and half to complete. It was adorned with statues of saints and martyrs and terminates on each bank with an imposing tower gateway.

In France during this time, a fine example of the early French Renaissance, the Pont Neuf, was being designed. It was the second stone bridge to be built in Paris and, although its design and construction did not represent a great leap forward in bridge building, it occupies a special place in Parisian hearts. It was designed by Jacques Androuet du Cerceau, and its two arms, which join the Ile de la Cité to the left and right bank of the Seine, represented a massive undertaking. Although all the arches are semicircular and not segmental, no two spans are alike, as they vary from 31 to 61 feet in span and also differ on the downstream and upstream sides of each arch, which were built on a skew of 10 percent. Du Cerceau wanted the bridge to be a true unencumbered thoroughfare bereft of any houses and shops. But the people of Paris demanded shops and houses and got their way in the end, which resulted in modification to the few short-span piers that had been constructed.

TOP: Pont Neuf, Paris, France (c1607).

ABOVE: The semi-circular arches of the Pont Neuf.

The Pont Neuf has stood now for 400 years and was the center of trade, and the principal access to the crowded island when it was built. The booths and stalls on the bridge became so popular that all sorts of traders used it, including booksellers, pastry cooks, jugglers, and peddlers. They crowded the roadway until there were some 200 stalls and booths packed into every niche along the pavement.

The longer left bank of the Pont Neuf was extensively reconstructed in 1850 to exactly the same details, after many years of repairs and attention to its poor foundations. The right bank with the shorter spans has been left intact. The entire bridge has been cleared of all stalls and booths and is used today as a road bridge.

The finest examples of late French Renaissance bridge building during the seventeenth century were the Pont Royal and Pont Marie bridges, which are still standing today. The Pont Royal was the first bridge in Paris to feature elliptical arches and the first to use an open caisson to provide a dry working place in the river bed. The foundations for the bridge piers were designed and constructed under the supervision of Francosi Romain, a preaching brother from Holland who was an expert in solving difficult foundation problems. The bridge architect, François Mansart, and the builder, Jacques Gabriel, called on Romain after they ran into foundation problems.

Romain introduced dredging in the preparation of the river bed for the caisson using a machine that he had developed. After excavations were finished the caisson was sunk to the bed, but the top was kept above the water level. The water was then pumped out and the masonry work of the pier was built inside the dry chamber. The five arch spans of the Pont Royal increase in span toward the center and, although the bridge has practically no ornamentation, it blends beautifully into its river setting.

TOP: Elliptical arches of Pont Royale in Paris (1687).

ABOVE: Pont Marie, Paris (seventeenth century).

The Renaissance brought improvement in both the art and science of bridge building. For the first time people began to regard bridges as civic works of art. The master bridge builder had to be an architect, structural theorist, and practical builder, all rolled into one.

The bridge that was without doubt the finest exhibition of engineering skills in this era was the slender, elliptical-arched bridge of Santa Trinità at Florence, designed by Bartolommeo Ammannati, in 1567. Many scholars are still mystified to this day as to how Ammannati arrived at such pleasing slender, curves to the arches.

The eighteenth century: The Age of Reason

In this period masonry arch construction reached perfection, owing to a momentous discovery by Jean Perronet and the innovative construction techniques of John Rennie. Just as the masonry arch reached its zenith 7,000 years after the first crude corbeled arch in Mesopotamia, it was to be threatened by a new building material—iron—and the timber truss, as the principal construction for bridges in the future.

This was the era when the civil engineering as a profession was born, when the first school of engineering was established in Paris at the Ecole de Paris during the reign of Louis XV. The director of the school was Jacques Gabriel, who had designed the Pont Royal. He was given the responsibility of collecting and assimilating all the information and knowledge there was on the science and history of bridges, buildings, roads, and canals.

With such a vast bank of collective knowledge, it was inevitable that building architecture and civil engineering should be separated into the two fields of expertise. It was suggested it was not possible for one man in his brief life to master the essentials of both subjects. Moreover, it became clear that the broad education received in civil engineering at the Corps des Ponts et Chaussées (or Bridge and Highway Corps) at the Ecole de Paris was not sufficient for the engineering of the large projects. More specialized training was needed in bridge engineering. In 1747 the first school of bridge engineering was founded in Paris at the historic Ecole des Ponts et Chaussées. The founder of the school was Trudiane, and the first teacher and Director was a brilliant young engineer named Jean-Rodolphe Perronet.

ABOVE: Rennie's "New London Bridge" under construction.

Jean Perronet has been called the "father of modern bridge engineering" for his inventive genius and design of the greatest masonry arch bridges of the century. In his hand the masonry arch reached perfection. The arch he chose was the curve of a segment of a circle of larger radius, instead of the familiar three-centered arch. To express the slenderness of the arch he raised the haunch of the arch considerably above the piers.

Perronet was the first person to realize that the horizontal thrust of the arch was carried through the spans to the abutments, and that the piers, in addition to carrying the vertical load, also had to resist the difference between adjacent span thrusts. He deduced that, if the arch spans were about equal and all the arches were in place before the centering was removed, the piers could be greatly reduced in size.

What remains of Perronet's great work? Only his last bridge, the glorious Pont de la Concorde in Paris, built when he was in his eighties. It is one of the most slender and daring stone-arch constructions ever built. "Even with modern analysis," suggests Professor James Finch, the author of *Engineering and Western Civilization*, "we could not further refine Perronet's design."

With France under the inspired leadership of Gabriel and then Perronet, the rest of Europe could only admire and copy these great advances in bridge building. In England a young Scotsman named John Rennie was making his mark, following in the footsteps the great French engineers. He was regarded as the natural successor to Perronet, who was a very old man when Rennie began his career. Rennie was a brilliant mathematician, a mechanical genius and pioneering civil engineer. In his early years he worked for James Watt to build the first steam-powered grinding mills at Abbey Mills in London, and later designed canals and drainage systems to drain the marshy fens of Lincolnshire.

Rennie built his first bridge in 1779 across the Tweed at Kelso. It was a modest affair with a pier-width-to-span ratio of 1:6 and with a conservative elliptical arch span. He picked up the theory of bridge design from textbooks and from studies and discussion about arches and voussoirs with his mentor Dr Robison of Edinburgh University. He designed bridges with a flat, level roadway and not the characteristic hump of most English bridges. It was radical departure from convention and was much admired by all the townsfolk, farmers, and traders who transported material and cattle across them.

TOP: "New London Bridge" was completed in 1831.

ABOVE: Rennie's bridge was rebuilt at Lake Havanasu, Arizona in 1972.

This bridge was a modest forerunner to the many famous bridges that Rennie went on to build—including Waterloo, Southwark, and New London Bridge. What was Rennie's contribution to the *science* of bridge building then? For Waterloo bridge, the centering for the arches was assembled on shore then floated out on barges into position. So well and efficiently did this system work that the framework for each span could be put in position in a week. This was a fast erection speed and, as a result, Rennie was able to halve bridge construction time.

So soundly were Rennie's bridges built that 40 years later Waterloo bridge had settled only 5 inches. Rennie's semi-elliptical arches, sound engineering methods and rapid assembly technique, together with Perronet's segmental arch, divided pier, and understanding of arch thrust, changed bridge design theory for all time.

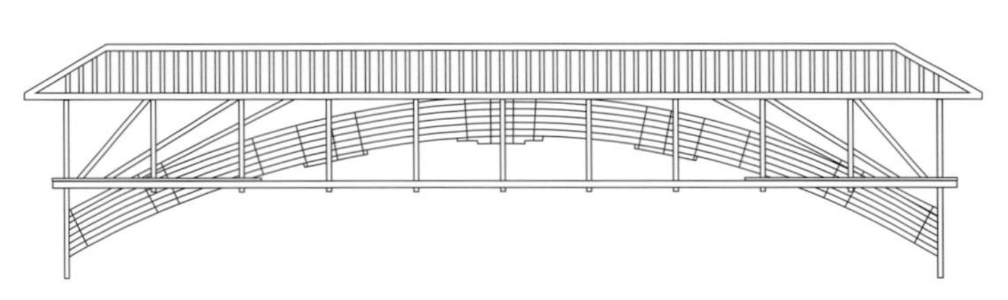

TOP: The covered bridge at Hartland, New Brunswick. As it proudly states, it is the world's longest covered bridge.

ABOVE: Illustration of the Wettingen bridge, Switzerland.

The carpenter bridges

America with her vast expansion of roads and waterways, following her commercial growth in the eighteenth century, was to become the home of the timber bridge in the nineteenth.

America had no tradition or history of building with stone, and so early bridge builders used the most plentiful and economical material available: wood. They produced some of the most remarkable timber bridge structures ever seen, but they were not the first to pioneer such structures. The Grubenmann brothers of Switzerland were the first to design a quasi-timber truss bridge in the eighteenth century. The Wettingen bridge over the Limmat just west of Zürich was considered their finest work. The bridge combines the arch-and-truss principle with seven oak beams bound closely together to form a catenary arch to which a timber truss was fixed. The span of the Wettingen was 309 feet and far exceeded the span of any other timber bridge at the time. Palmer, Wernwag, and Burr—the so called carpenter-bridge builders of North America, who designed more by intuition than by calculation—developed the truss arch to span farther than any other wooden construction. This was the last of the three basic bridge forms to be discovered. The first man who made the truss arch bridge a success in America and who patented his truss design was Timothy Palmer, a New England Yankee. In 1792 he built a bridge consisting of two trussed arches over the Merrimac; similar to one of Palladio's truss designs, except the arch was the dominant supporting structure.

Palmer's "Permanent Bridge" over the Schuylkill, built in 1806, was his most celebrated bridge. When the bridge was finished the president of the bridge company suggested that it would be a good idea to cover it to preserve the timber from rot and decay in the future. Palmer went further than that and timbered the sides as well, completely enclosing the bridge. Thus America's distinctive covered bridge was established. By enclosing the bridge it stopped snow getting in and piling up on the deck, thus causing it to collapse from the extra load.

Wernwag was a German immigrant from Pennsylvania, who built 29 truss-type bridges in his lifetime. His designs integrated the arch and truss into one composite structure rather more successfully than Palmer's.

Wernwag's famous bridge was the "Colossus" over the Schuylkill just upstream from Palmer's "Permanent Bridge" and comprised two pairs of parallel arches, linked by a framing truss, which carried the roadway. The truss itself was acting as bracing reinforcement and consisted of heavy verticals and light diagonals. The diagonal elements were remarkable because they were iron rods, and were the first iron rods to be used in a long-span bridge. In its day the Colossus was the longest wooden bridge in America, having a clear span of 304

feet. Fire destroyed the bridge in 1838. It was replaced with Charles Ellet's pioneering suspension bridge.

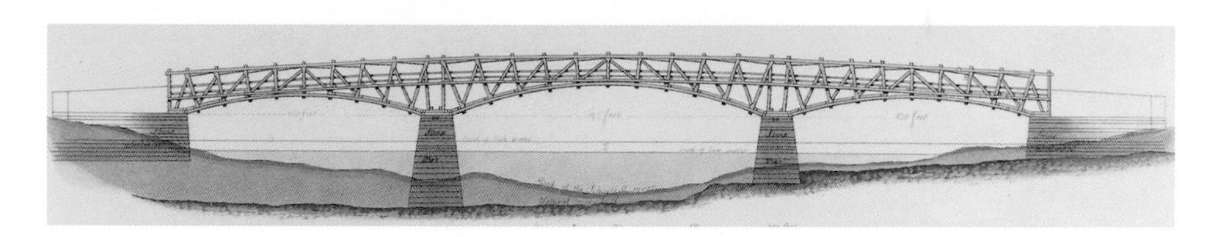

Theodore Burr was the most famous of the illustrious triumvirate. Burr developed a timber truss design based on the simple king-and-queen-post truss of Palladio. He came closest to building the first true truss bridge, but it proved unstable under moving loads. Burr then strengthened the truss with an arch. It was significant that here the arch was added to the truss rather than the other way round.

ABOVE: Palmer's "Permanent Bridge" over the Schuykll River, USA.

Burr's arch trusses were quick to assemble and modest in cost to build, and for a time they were the most popular timber bridge form in America. By 1820 the truss principle had been well explored and, although the design theory was not understood, in practice it had been tested to the limit.

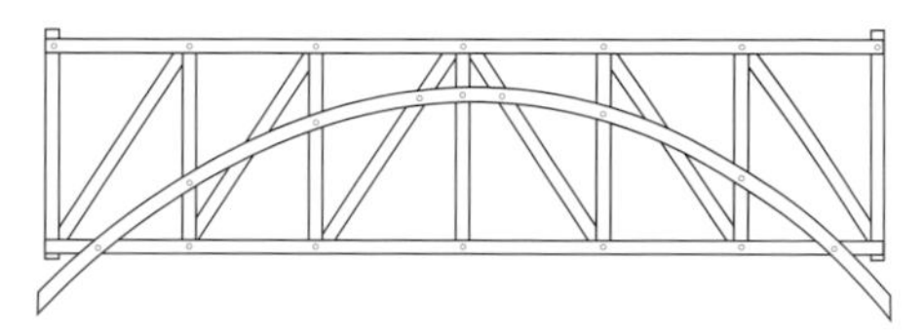

The Burr arch/truss (1815)

It was left to Ithiel Town to develop and build the first true truss bridge, which he patented and called the Town Lattice. It was a true truss because it was free from arch action and any horizontal thrust. Simple to build, it could be nailed together in a few days and was cheap compared with other options. Town promoted his bridge with the slogan "built by the mile and cut by the yard."

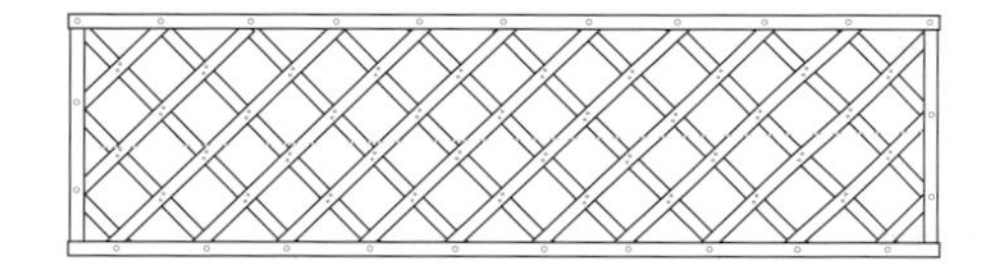

The Town lattice truss (1820)

With the arrival of the railroad in the US, bridge building continued to develop along two separate ways. One school continued to evolve stronger and leaner timber truss structures, while the other experimented with cast and wrought iron.

The first patent truss to incorporate iron into a timber structure was the Howe Truss—named after William Howe, who pioneered the development of truss bridges in the US. It had top and bottom chords and diagonal bracing in timber and the vertical members made of iron rods in tension. This basic design was, with modification, to last right into the next century.

The first fully designed truss was the Pratt Truss, which reversed the forces of the Howe Truss by putting the vertical timber members in compression and the iron diagonal members in tension. In 1847, the Whipple truss (named after the American civil engineer Squire Whipple) was the first all-iron truss—a bowstring truss—with the top chord and vertical compression members made from cast iron and the bottom chord and diagonal bracing members made from wrought iron.

Later Fink, Bollman, Bow, and Haupt in the US, along with Cullman and Warren in Europe, developed the truss to a fine art, incorporating wire-strand cable, timber, and iron to form lightweight, strong bridges to carry railroads.

BELOW: "Bollman Truss Bridge" (1869), a famous iron bridge in Savage, Maryland, USA.

2

A pictorial history of modern bridge building—the last 200 years

Bridge development in the nineteenth and twentieth centuries

ABOVE: The steel-arched St Louis bridge, USA (1874).

OPPOSITE: The cable-stayed Coatzacoalas 11 Bridge, Mexico (1984).

The Industrial Revolution, which began in Britain at the end of the eighteenth century, gradually spread throughout the world and brought with it huge changes in all aspects of everyday living. New forms of bulk transportation, by canal and rail, were developed to keep pace with the increasing exploitation of coal and the manufacture of textiles and pottery. Coal fueled the hot furnaces to provide the high temperatures to smelt iron. Henry Bessemer invented a method to produce crude steel alloy by blowing hot air over smelted iron. Siemens and Martins refined this process further to produce the low-carbon steels of today.

High temperature was also essential in the production of cement, which John Aspen discovered by burning limestone and clay on his kitchen stove in Leeds, England, in 1824. Wood and stone were gradually replaced by cast-iron and wrought-iron construction, which in turn was replaced by first steel and then concrete, the two primary materials of bridge building in the twentieth century.

"Old London Bridge" fifteenth century; stone pointed arch.

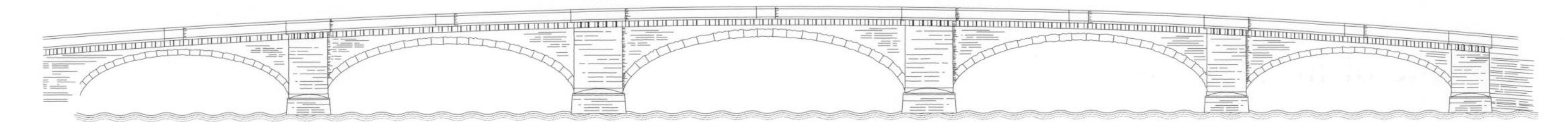

"New London Bridge" 1831; stone segmental arch.

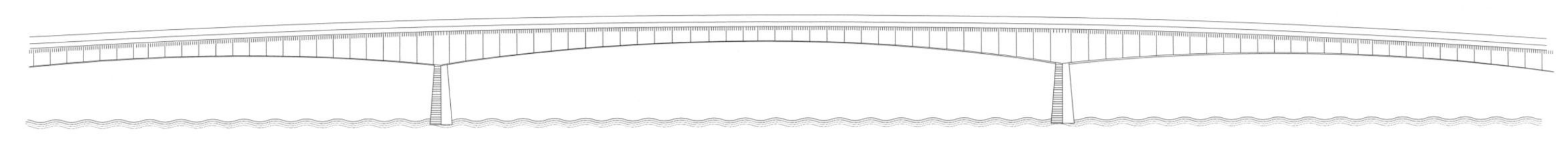

"London Bridge" 1968; prestressed concrete flat arch.

ABOVE: Progress of arch construction from the fifteenth to the twentieth century.

BELOW: Brooklyn Bridge, New York, 1993.

BOTTOM: Brooklyn Bridge, New York, 1858.

Growing towns and expanding cities demanded continuous improvement and extension of the road, canal, and railroad infrastructure. The machine age introduced the steam engine, the internal-combustion engine, factory production lines, domestic appliances, electricity, gas, processed food, and the tractor.

Faster assembly of bridge building was essential, and this meant prefabricating lightweight but tough bridge components. The heavy steam engines and longer goods train imposed larger stresses on bridge structures than ever before. Bridges had to be stronger and more rigid in construction and yet had be faster to assemble to keep pace with progress. Connections had to be stronger and more efficient. The common nut and bolt were replaced by the rivet, which was replaced by the high-strength friction-grip bolt and the welded connection.

When the automobile arrived, it resulted in a road network that eventually crisscrossed the entire countryside from town to city, over mountain ranges, valleys, streams, rivers, estuaries, and seas. Even bigger and better bridges were now needed to connect islands to the mainland and countries to continents, to open up major trading routes. The continuous search for and development of high-strength materials such as steel, concrete, carbon fiber, and aramids have today combined with sophisticated computer analysis and dynamic testing of bridge structures against earthquakes, hurricane wind, and tidal flows to enable bridges to span even farther. In the last two centuries the bridge span has leapt from 350 to over 6,000 feet. This is the age of the mighty suspension bridges, the elegant cable-stayed

bridges, the steel-arch truss, the glued segmental and cantilever box-girder bridges.

The key events and achievements in this frenzied activity of bridge building are highlighted in pictorial form to illustrate the rapid pace of change and the many bridge ideas that were advanced. In the last two centuries more bridges were built than in the entire history of bridge building prior to that!

The age of iron (1775–1880)

Of all the materials used in bridge construction—stone, wood, brick, steel, and concrete—iron was used for the shortest time. Cast iron was first smelted from iron ore successfully by Dud Dudley in 1619. It was another century before Abraham Darby devised a method to economically smelt iron in large quantities. However, the brittle quality of cast iron made it safe to use only in compression in the form of an arch.

Wrought iron, which replaced cast iron many years later, was a ductile material that could carry tension. It was produced in large quantities after 1783 when Henry Cort developed a puddling furnace process to drive impurities out of pig iron.

But iron bridges suffered some of the worst failures and disasters in the history of bridge building. The vibration and dynamic loading from a heavy steam locomotive and goods wagons create cyclic stress patterns on the bridge structure as the wheels roll over the bridge, going from zero load to full load, then back to zero. Over a period of time these stress patterns can lead to brittle failure and fatigue in cast iron and wrought iron.

In one year alone in the US, as many as one in every four iron and timber bridges had suffered a serious flaw or had collapsed. Rigorous design codes, independent checking, and new bridge-building procedures were drawn up, but it was not soon enough to avert the worst disaster in iron-bridge history over the Tay Estuary in Scotland in 1879. It marked the end of the iron bridge for good.

ABOVE: Iron Bridge, Coalbrookdale, England (1779).

LEFT: Buildwas Bridge, Coalbrookdale, England (1796).

Landmarks of the age of iron

- 1779 **Coalbrookdale, England, the first cast-iron bridge, designed as an arch structure by Thomas Pritchard for its owner and builder, Abraham Darby III.**
- 1796 **Buildwas Bridge, the second cast-iron bridge built in Coalbrookdale, England, designed by Thomas Telford, used only half the weight of cast iron of Coalbrookdale.**
- 1807 **James Finlay builds the first elemental suspension bridge in wrought iron, the Chain Bridge, over the Potomac in Washington, DC.**
- 1821 **Guinless Bridge, England, George Stephenson's wrought-iron "lentilcular" girder bridge for the Stockton to Darlington Railway.**
- 1826 **Menai Straits Bridge, Wales, famous eyebar, wrought-iron chain-suspension bridge over the Menai Straits, by Thomas Telford.**
- 1834 **The Fribourg Bridge, Switzerland, the world's longest iron suspension bridge.**
- 1841 **Whipple patents the cast-iron "bowstring" truss bridge.**
- 1846 **Wheeling Suspension Bridge, USA, Charles Ellet's record-breaking, 1,000-foot-span, iron-wire suspension bridge.**
- 1850 **Britannia Bridge (Wales), first box-girder bridge concept, built in wrought iron by Robert Stephenson, son of George Stephenson.**
- 1853 **Murphy designs a wrought-iron Whipple truss, with pin connections.**
- 1858 **Royal Albert Bridge, Saltash, in the southwest of England: Isambard Kingdom Brunel's famous tubular-iron bridge over the Tamar.**
- 1876 **The Ashtabula Bridge disaster in USA: 65 people die when this iron, modified Howe truss collapses plunging train and passengers into the deep river gorge below.**
- 1879 **The Tay Bridge disaster, Dundee, Scotland, where a passenger train with 75 people on board plunges into the Tay estuary, as the supporting wrought iron girders collapse in high winds.**

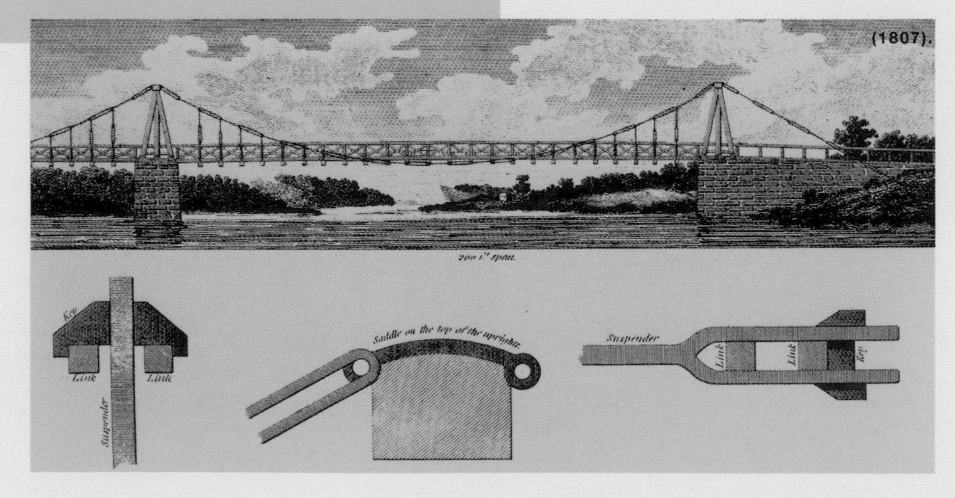

FINLAYS "CHAIN SUSPENSION BRIDGE" (1807).

FRIBOURG BRIDGE, SWITZERLAND (1834).

FABRICATING THE BOX SECTIONS OF THE BRITANNIA BRIDGE.

BRITANNIA BRIDGE, WALES (1850).

NIAGARA BRIDGE, USA (1855).

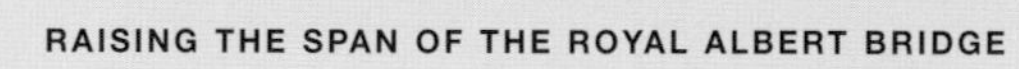

RAISING THE SPAN OF THE ROYAL ALBERT BRIDGE

BAYONNE BRIDGE, USA (1931).

The Age of Steel

Out of the chrysalis of cast iron and wrought iron in the nineteenth century evolved steel—a stronger and highly ductile alloy that was not susceptible to brittle failure. In 1856 Henry Bessemer patented a method for reducing the carbon content of pig iron, to produce the first crude steel. Later inclusion of small amounts of nickel, silicone and manganese alloys with pig iron and a hot rolling process produced the ductile, high strength, modern steels that were used for building the great bridges of the twentieth century. Today steel wire strands and hot rolled steel sections can be fabricated to form lightweight trusses, slender arches, great steel suspension bridges and graceful cable stay bridges. In the last 100 years steel has been the primary material in bridge building, and the material used in building the record-breaking spans of the Brooklyn, Firth of Forth, Quebec, George Washington, Golden Gate, Verazzano Narrows, Humber, Great Storebelt, and Akashi Kaikyo Bridges.

ABOVE: Sunrise and making ready to spin the cables.

RIGHT: The steel truss approach spans of the Mackinac Bridge.

FAR RIGHT: Assembling the steel sections of a truss.

ROYAL ALBERT BRIDGE, SALTASH, ENGLAND (1858).

WHEELING BRIDGE, USA (1846).

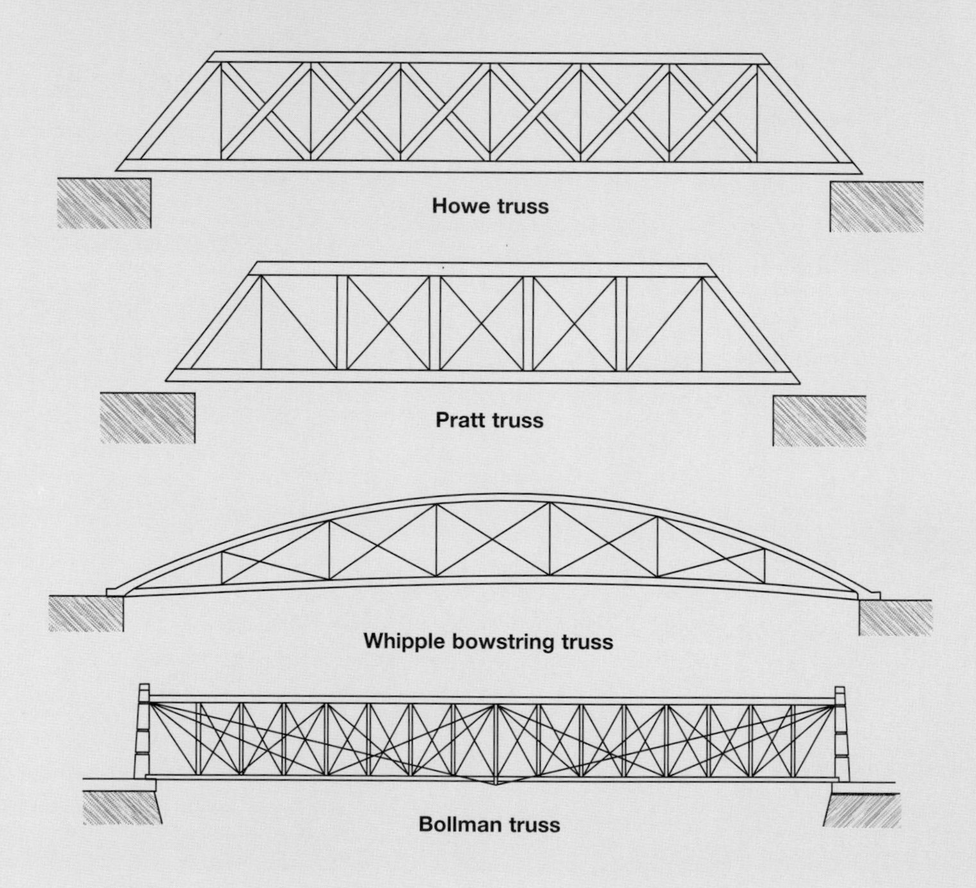

ABOVE: American truss bridges.

The arrival of steel

Steel is simply a refined iron from which carbon and other impurities have been driven off. Techniques for making steel were allegedly known in China in 200 BC and in India in 500 BC. But the process was very slow and laborious and, after a great deal of time and energy, only minute amounts were produced. It was very expensive, so until the nineteenth century it was used only for edging tools and weapons.

In 1856, Henry Bessemer, the English inventor and engineer, developed a process for bulk steel production by blowing air through molten iron to burn off the impurities. It was followed by the open-hearth method patented by Charles Siemens and Pierre Emile Martins in Birmingham, England, in 1867, which is the basis for modern steel manufacture today.

It took a while for steel to supersede iron, because it was expensive to manufacture. But, when the world price of steel dropped by 75 percent in 1880, it was suddenly competitive with iron. It had vastly superior qualities, both in compression and tension—it was ductile and not brittle like iron, and was much stronger. It could be rolled, cast, or even drawn, to form rivets, wires, tubes, and girders.

The age of steel opened the door to tremendous advances in long-span bridge-building technology. The first bridges to exploit this new material were in America, where the steel arch, the steel truss, and the wire-rope suspension bridges were pioneered. Later Britain led the world in the cantilever-truss bridge and the steel box-girder bridge deck.

Let us now map the historical progress of the principal bridges in steel covering a period from 1880 to the present time in the following order: the steel truss-arch bridge, the cantilever-truss bridge, the suspension bridge, and steel-plate-girder and box-girder bridges.

The steel truss-arch bridge When steel prices dropped in the 1870s and 1880s the first important bridges to use steel were all in the United States. The arches of St Louis Bridge over the Mississippi and the five Whipple trusses of the Glasgow Bridge over the Missouri were the first to incorporate steel in truss construction. St Louis, near the confluence of the Missouri and Mississippi, was the most important town in the Midwest, and the focal point of north–south river traffic and east–west overland routes.

BACKGROUND: Inside the steel arch of the Hellgate Bridge.

Landmarks of the steel arch

- 1874 **The St Louis Bridge, USA—James Eades builds the first triple-arch steel bridge.**
- 1884 **The Garabit Viaduct, St Flour, France—Gustav Eiffel's truss arch in wrought iron was the prototype for future steel truss construction. Eiffel would have preferred steel but chose wrought iron because it was more reliable in quality and cheaper.**
- 1916 **The Hell Gate Bridge, New York—the first 1,000-foot steel-arch span in the world was designed by Gustav Lindenthal.**
- 1931 **The Bayonne Bridge, New York—the first bridge to be built with a cheaper carbon manganese steel, rather than nickel steel, and which is the composition of most modern steel.** ***{see page 35}*****.**
- 1932 **Sydney Harbour Bridge, Sydney—this famous steel arch was built using 50,000 tons of nickel steel. Its design was based on the Hell Gate Bridge.**
- 1978 **New River Gorge Bridge, West Virginia—currently the world's longest steel-arch span.**

HENRY HUDSON BRIDGE, USA (1938).

GARABIT VIADUCT, ST FLOUR, FRANCE (1884).

HELLGATE BRIDGE, NEW YORK, USA (1916).

SYDNEY HARBOUR BRIDGE, AUSTRALIA (1932).

NEW RIVER GORGE BRIDGE, USA (1978).

The cantilever truss Arch bridges had been constructed for many centuries in stone, then iron, and steel when it became available. Steel made it possible to build long-span trusses farther than cast iron, without any increase in the dead weight. Consequently, it made cantilever long-span truss construction viable over wide estuaries. The first and most significant cantilever truss bridge to be built was the railroad bridge over the Firth of Forth near Edinburgh, Scotland, in 1890. The cantilever truss was rapidly adopted for the building of many US railroad bridges until the collapse of the Quebec Bridge in 1907.

ABOVE: Forth Rail Bridge, Scotland (1890).

Landmarks

- 1886 **The Fraser River Bridge, Canada—believed to be the first balanced cantilever truss bridge to be built. All the truss piers, links, and lower chord members were fabricated from Siemens-Martin steel. It was dismantled in 1910.**
- 1890 **The Forth Rail Bridge, Edinburgh, Scotland—the world's longest spanning bridge at 1709 ft, when it was finished.**
- 1891 **The Cincinnati Newport Bridge, Cincinnati, USA—with its long, through, cantilever spans and short-truss spans—was the prototype of many railroad bridges in the USA.**
- 1902 **The Viaur Viaduct, France—this railroad bridge between Toulouse and Lyons was an elegant variation of the balanced cantilever, with no suspended section between the two cantilever arms.**
- 1918 **The Quebec Bridge, Canada—completion of the second Quebec bridge, the world's longest cantilever span.**
- 1927 **Carquinez Bridge—the last of the long cantilever truss bridges to be built in the US, although a second, identical, bridge was built alongside it in 1958 to increase traffic flow.**

THOUSAND ISLANDS BRIDGE—THE CANTILEVER TRUSS SPAN—CANADA, USA (1938)

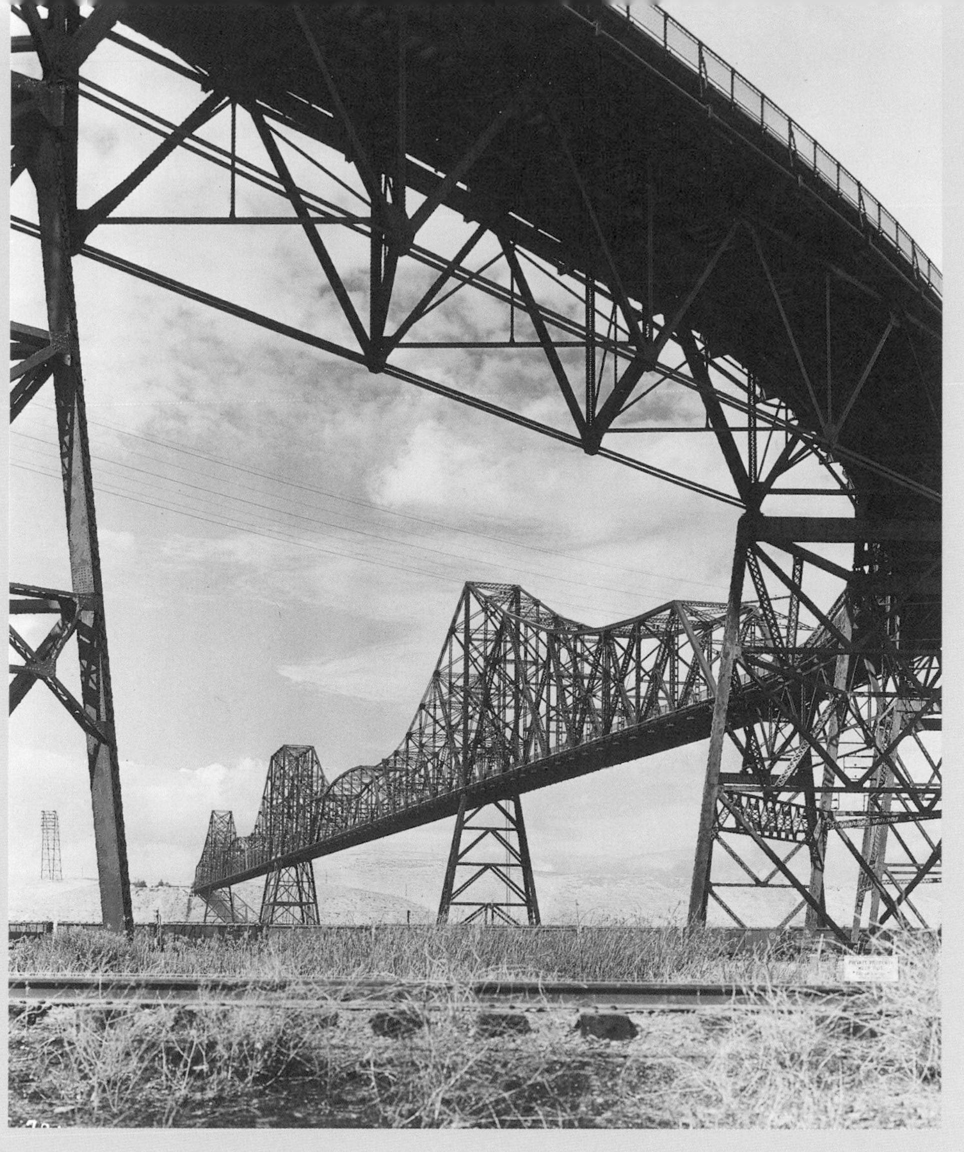

CARQUINEZ BRIDGE, USA (1927).

BRIDGE OF THE GODS, USA (1926).

POUGHKEEPSIE RAIL BRIDGE, USA (1888).

QUEBEC BRIDGE, CANADA (1918).

VIAUR VIADUCT, FRANCE (C1900).

The suspension bridge The early pioneers of chain suspension bridges were James Finlay, Thomas Telford, Samuel Brown, and Marc Seguin, but they had only cast and wrought iron available in the building of their early suspension bridges. It was not until Charles Ellet's Wheeling Bridge had shown the potential of the wire suspension principle using wrought iron that the concept was universally adopted. Undoubtedly, the greatest exponent of early wire suspension construction and strand-spinning technology was John Roebling. His Brooklyn Bridge was the first to use steel for the wires of suspension cables.

Suspension bridges are capable of huge spans, bridging wide river estuaries and deep valleys, and have been vital in establishing road networks across a country. They have held the record for the longest span almost unchallenged from 1826 to the present day—a record interrupted only between 1890 and 1928, when the cantilever truss held the record.

RIGHT: David Steinmann on the Brooklyn Bridge, after it was strengthened in 1953.

Landmarks—suspension bridges

- 1883 **Brooklyn Bridge, USA—following the completion of the Wheeling suspension bridge pioneered by Charles Ellet; John Roebling went on to design the Brooklyn Bridge, the first steel-wire suspension bridge in the world.**
- 1931 **George Washington, USA—the heaviest suspension bridge to use parallel wire cables rather than rope-strand cables, and had the longest span in the world for nearly a decade.**
- 1940 **Tacoma Narrows, USA—the second Tacoma Narrows Bridge over Puget Sound in Washington State. This bridge, rebuilt after the collapse of the first bridge, with a deep stiffening truss deck, set the trend for future suspension bridge design in America.**
- 1957 **Mackinac, USA—Fondly referred to as "Big Mac" is the longest overall suspension bridge in America.**
- 1964 **Verazzano Bridge, USA—the last big suspension bridge to be built in America, and also held the record for the longest span until 1981.**
- 1967 **Severn Bridge, England—the first bridge to have a slim, aerodynamic bridge deck, eliminating the need for deep stiffening trusses like those of the American suspension bridges. It set the trend for future suspension-bridge construction.**
- 1981 **Humber Bridge, England—the longest span in the world when it was completed, with supporting strands that were inclined in zigzag fashion rather than the parallel arrangement preferred by the Americans.**
- 1998 **Storebelt and the East Bridge, Denmark—Storebelt is now substantially complete, and is the longest bridge in Europe. For a short while the main span of the East Bridge held the record for the longest span in the world.**
- 1998 **Akashi Kaikyo, Japan—is one of a family of long-span bridges linking the islands of Honshu and Shikoku, now well under construction. Its main span of 6,528 feet makes it the longest span in the world.**

BROOKLYN BRIDGE, USA (1883).

LION'S GATE BRIDGE, VANCOUVER, BRITISH COLUMBIA, CANADA

CABLE ANCHORAGES OF THE THOUSAND ISLANDS SUSPENSION BRIDGE, CANADA, USA (1938).

OAKLAND BAY BRIDGE, SAN FRANCISCO, USA (1932).

ST JOHN'S BRIDGE, OREGON, USA (1931).

GEORGE WASHINGTON BRIDGE, USA (1931).

FROZEN WATERS UNDER THE MACKINAC BRIDGE.

TAGUS BRIDGE, LISBON, PORTUGAL (1966).

MACKINAC BRIDGE, USA (1957).

HUMBER BRIDGE, ENGLAND (1981).

ELEVATION OF THE TAGUS BRIDGE.

AKASHI KAIKYO BRIDGE, JAPAN (1998).

EAST BRIDGE, GREAT STOREBELT, DENMARK (1998).

VERAZZANO NARROWS BRIDGE, USA (1964).

Steel-plate-girder and box-girder bridges Since the development of steel and the I-beam, many beam bridges were built using a group of beams in parallel, which were interconnected at the top to form a roadway. They were quick to assemble but they were practical over only relatively short spans for rail and road viaducts. The riveted girder I-beam was later superseded by the welded and friction-grip bolted beam.

Relatively long spans were not efficient, however, because the depth of the beam needed could became excessive. To counter this, web plate stiffeners were added at close intervals to prevent buckling of the beam. Another solution was to make the beam into a hollow box, which was very rigid. In this way the depth of the beam could be reduced and material could be saved.

BELOW: Elbe Bridge, Germany—steel plate girder (1936); Bonn-Buel Bridge, Germany—steel plate girder (1948).

The steel box-girder beams were quick to fabricate and easy to transport. Their relatively shallow depth meant that high approaches were not necessary. Most of this pioneering work was carried out during and after World War II, when there was a huge demand for fast and efficient bridge building for spans of up to 1,500 feet. A major rebuilding program in Germany witnessed the construction of many steel box-girder and concrete box-girder bridges in the 1950s and 1960s. For spans greater than 1,500 feet the suspension and cable-stay bridge are generally more economical to build.

In the 1970s the world's attention was focused on the collapse of four steel box-girder bridges under construction. The four bridges were in Vienna over the Danube, in Milford Haven in West Wales (when four people were killed), a bridge over the Rhine in Germany, and the West Gate bridge in Melbourne, Australia, over the Lower Yarra River. By far the worst collapse was on the West Gate bridge, a single-cable-stay structure with a continuous box-girder deck. A cantilever section 376 feet long and weighing 1,200 tons buckled and crashed off the pier support onto some site huts 120 feet below, where workmen and engineers were having their lunch. Thirty-five people were killed in the tragedy. After that, further construction of steel box-girder decks was halted until better design standards, new site checking procedures, and a fabrication specification was internationally agreed.

Landmarks

- 1936 **Elbe Bridge—one of the early plate-girder bridges on the German autobahn.**
- 1948 **Bonn Beuel—a later development of the plate girder into a flat arch, to reduce material weight.**
- 1952 **Cologne Deutz Bridge—first slender, steel box-girder bridge in the world.**
- 1970s **Failure of box girders at Milford Haven in West Wales and West Gate Bridge in Australia stopped further development of the steel box-girder bridge decks.**

KOBLENZ BOX GIRDER BRIDGE COLLAPSE, GERMANY.

BONN SOUTH RHINE BRIDGE IN GERMANY—STEEL BOX GIRDER (1967).

BOX GIRDER BRIDGES, A FEATURE OF MANY HIGHWAY INTERCHANGES.

Concrete and the arch (1900) Although engineers took longer to realize concrete's true potential as a material, it is used everywhere today in a vast array of bridges and building applications. Concrete is a brittle material just like stone, good in compression but not in tension, so if it starts to bend or twist it will crack. Concrete has to be reinforced with steel to give it ductility, so naturally its emergence followed the development of steel.

In 1824 Joseph Aspen made a crude cement from burning a mixture of clay and limestone at high temperature. The clinker that was formed was ground into a powder, and when this was mixed with water it reacted chemically to harden back into a rock. Nowadays, cement is usually combined with sand, stones, and water to create concrete, which remains fluid and plastic for a period of time, before it begins to set and eventually hardens. You can pour and place concrete into molds or forms while it is fluid, to create bridge beams, arch spans, support piers—in fact a variety of structural shapes. This gives concrete special qualities as a material, and scope for bold and imaginative bridge ideas.

ABOVE: Glenfinnan Viaduct, Scotland (1898).

François Hennebique was the first to understand the theory and practical use of steel reinforcement in concrete, but it was Robert Maillart (1872–1940) who was first to pioneer and build bridges with reinforced concrete. Eugene Freyssinet, Maillart's contemporary, was also keen to experiment with concrete structures. He went on to discover the art of prestressing and gave the bridge industry one of the most efficient methods of bridge deck construction in the world. Both these men were great engineers and champions of concrete bridges. What they achieved set the trend for future developments in concrete bridges—precast bridge beams, concrete arches, and box-girder and segmental cantilever construction. Concrete box-girder bridge decks are incorporated into many modern cable-stay and suspension bridges.

Hans Wittfoht, Jean Muller and the contractors Polensky and Zöllner, and Campenon Bernard, were responsible for building the first segmental and cast in place concrete box-girder bridges in the world. It is a technique that is used by many bridge builders across the world today. The box-girder span can be precast as segments or cast in place using a traveling formwork system. They can be built as a balanced cantilever each side of a pier or launched from one span to the next.

Concrete has been used in building most of the world's longest bridges. The relative cheapness of concrete compared with steel, the ability to rapidly precast or form prestressed beams of standard lengths, to bridge short spans over low-level trestle-type supports, had made concrete economically attractive. Lake Pontchartrain Bridge, a precast-concrete, segmental, box-girder bridge in Louisiana, is the longest bridge in the US with an overall length of 23 miles.

The Concrete Arch

- 1905 **Glenfinnan Viaduct, Scotland—the first concrete arch bridge to be built in England.**
- 1905 **Tavanasa Bridge, Switzerland—a breakthrough in the stiffened-arch slab.**
- 1922 **St Pierre de Vouvray, France—early concrete bowstring arch of Freyssinet.**
- 1930 **Salgina Gorge Bridge, Switzerland—one of the most aesthetic arch spans of Maillart.**
- 1930 **Plougastel Bridge, France—unique construction concept which used prestressing for the first time.**
- 1936 **Alsea Bay Bridge, USA—completion of Conde McCullough's finest "art deco" bridge in concrete.**
- 1964 **Gladesville, Australia—use of precast, prestressed, segmental construction for the arch span. Also in 1964, KRK (Croatia)—the longest concrete arch span in the world.**

GLADESVILLE BRIDGE, AUSTRALIA (1964).

PLOUGASTEL BRIDGE, FRANCE (1930).

KRK BRIDGE, CROATIA (1964).

SALGINA GORGE BRIDGE, SWITZERLAND (1930).

EARLY PICTURE OF THE SALGINA GORGE BRIDGE, SWITZERLAND

SCHWANDBACH BRIDGE, SWITZERLAND (1933).

Concrete box girders In the 1950s and 1960s many freeway bridges and viaducts were built in Europe and USA using concrete box-girder construction. Some were precast segmental construction; some were cast in place.

FOLLOWING FOUR PAGE SPREAD:

General Belgrano Bridge over the Parana River in Argentina (1973)—cable stay navigation span, concrete box girder approach spans.

Landmarks

- 1952 **Shelton Road Bridge, USA—first match-cast, glued-segmental, box-girder construction in the world, developed by Jean Muller.**
- 1956 **Lake Pontchartrain Bridge, USA—the second longest bridge in the world. It is a precast, segmental, box-girder bridge with 2,700 spans and runs for 23 miles across Lake Pontchartrain near New Orleans. The second, identical, bridge, was built alongside the original one in 1969 and was just a little longer.**
- 1964 **Krahnenberg Bridge**
- 1972 **Medway Bridge, England—the first European river bridge to be built using concrete box-girder construction.**

KOCHERTAL VIADUCT, GERMANY (1979).

KYLESKU BRIDGE, SCOTLAND (1984).

SECOND CHOISY BRIDGE, PARIS, FRANCE (1988).

ELEVATED WIDE BEAM VIADUCT IN HANOVER, GERMANY (1958).

OOSTERSCHELDE BRIDGE, NETHERLANDS (1965).

PONT ANNET, FRANCE—A PRESTRESSED BEAM BRIDGE (1947).

SALLINGSUND BRIDGE, DENMARK (1974).

CONFEDERATION BRIDGE, PRINCE EDWARD ISLAND, CANADA (1997).

VENTABREN VIADUCT, FRANCE (1998).

NORTH BRIDGE, DÜSSELDORF, GERMANY (1956).

TARTARA BRIDGE IN JAPAN HOLDS THE RECORD FOR THE LONGEST CABLE STAY SPAN (1998).

Cable-stay bridges Cable stays are an adaptation of the early rope bridges and of guy ropes for securing tent structures and the masts of sailing ships. When very rigid, trapezoidal, box-girder bridge decks were developed for suspension bridges, it allowed a single plane of stays to support the bridge deck directly. This meant that fewer cables were needed than a conventional suspension system, and there was no need for anchorages, so it was cheaper to construct. Cost and time have always been the principal motivators for change and innovation in bridge engineering.

The first modern cable-stay bridges were pioneered by German engineers just after World War II, led by Fritz Leonhardt, René Walter, and Jorg Schlaich. The cable-stay bridge is probably the most visually pleasing of all modern long-span bridge forms. In recent times the development of the cable-stay and box-girder bridge deck has continued with the work of the Swedish engineers COWI consult; bridge engineers Carlos Fernandez Casado of Spain; R. Greisch of Belgium; Jean Muller International; Sogelerg; EEG; and Michel Virlogeux of France.

TING KAU BRIDGE, HONG KONG (1997).

Cable-stay history

- 1955 **Stormstrund, Norway—the first cable-stay bridge.**
- 1956 **North Bridge, Düsseldorf—early harp arrangement for a family of cable-stay bridges over the Rhine. It was the prototype for many cable-stay bridges.**
- 1959 **Severins Bridge, Germany—the first to adopt an A-frame tower and the first bridge to use a fan configuration for the stays—a very efficient bridge form.**
- 1962 **Lake Maracaibo Bridge, Venezuela—an unusual composite cable-stay and concrete-frame support structures for a bridge built in Venezuela, using local labor.**
- 1962 **Nordelbe Bridge, Hamburg—the first bridge to use a single plane of cables: the deck was a stiffened rectangular box girder.**
- 1966 **Wye Bridge, England—a single cable stay from the mast supports the continuous steel box-girder bridge deck. Erskine Bridge built in 1971 was a better example of this construction.**
- 1974 **Brotonne Bridge, France—the first cable-stay bridge to use a precast-concrete box-girder deck and a single plane of cable stays.**
- 1984 **Coatzacoalcos II Bridge, Mexico—elegant pier and mast tower combining the rigidity of the A-frame with the economy of a single foundation.**
- 1995 **Pont de Normandie, France—breakthrough in the design of very long cable-stay spans.**

SEVERINS BRIDGE, COLOGNE, GERMANY (1959).

GANTER BRIDGE, SWITZERLAND (1980).

PONT DE NORMANDIE, FRANCE (1994).

ERSKINE BRIDGE, SCOTLAND (1974).

3

What makes a bridge stand up

ABOVE: Closing the main span of Ting Kau Bridge in Hong Kong.

OPPOSITE: Kurushima suspension bridge under construction in Japan.

In understanding how bridges work there are two important structural terms to recognize—tension and compression. All structures—whether a house, a skyscraper, a dome, an arch, or a suspension bridge—are either in tension or compression. If you can understand how tension and compression forces work, then you will begin to appreciate how bridges are able to span.

What are tension and compression, and how can we recognize them? We may not be able to "feel" the forces in a bridge structure, but we can recognize the effects of tension and compression quite easily. For instance, if we pull on a rope or a string we say that we have put the string or rope in "tension." If you push down on a wooden post or hammer a nail into wood, you are putting the post or the nail into "compression." To find out how it "feels" to be in tension, hold the knob of a closed door and pull on it. Your arm is put in tension: it is being stretched. If you want to feel compression, push with your arm stretched against the door. Your arm is now in compression.

In "tension," we are stretching the structure and trying to lengthen it. By pushing down or "compressing" a structure, we are trying to shorten it. The amount of lengthening or shortening cannot be seen by the naked eye.

Tension is—pulling a string or rubber band apart or trying to pull open a door that has jammed. **Compression** is—pressing down on a cardboard box, squeezing a sponge or pushing hard against a door to stop it opening.

As a bridge bends under the weight of the load it is supporting, some parts of the structure may go into tension and others go into compression. For instance, an arch goes only into compression, while a suspension cable stretches and undergoes tension—but a truss or beam will undergo both tension and compression at the same time. Materials for bridge building that are useful in compression are stone, brick, and plain concrete. They are brittle and will crack if they are stretched or bent. Materials that are good in tension are rope, bamboo, and wire. They can be stretched, but, because they are not very rigid, they will not hold their shape under compression. However, highly engineered modern materials like steel and reinforced concrete are good in both tension and compression. Successful bridge builders through the ages had to design bridges to suit the building materials that were available to them at the time.

Understanding the tension and compression properties of materials, and how to control these forces in a bridge, helps to determine the type of bridge to be built and the length of its span.

What is Reinforced and Prestressed Concrete? Concrete is an artificial rock comprising sand, stones, and water, which when mixed with cement hardens to full strength in about four weeks. It is produced in ready-mix concrete plants and sent to the bridge works in a concrete wagon or truck, and then pumped into position while it is still fluid and workable. Sometimes it is poured into molds or forms in a casting yard to make precast concrete beams and sections, which are then transported to the site of the bridge works.

When water is added to cement, a chemical reaction takes place, which converts the very fine cement powder into a hard, crystalline rock material. The process is called hydration, and it is a complex chemical reaction. Water is also needed to lubricate the concrete mix to make it workable and easy to pour and compact into molds or forms.

Concrete is very good in compression, but, if you try to bend it or stretch it—putting in into tension—it will crack and fail. For concrete to be ductile, so that it can bend and recover and be capable of resisting tension forces, it has to be reinforced with steel rods or bars—materials called, predictably, "reinforcement." The reinforcement is positioned close to the face of the concrete that is in tension to prevent it cracking and to resist the tension stresses acting on it. Reinforcement placed in concrete will not rust, despite its being covered in wet concrete, because the pH of cement is so alkaline that it inhibits any rust and corrosion forces developing.

Prestressing concrete is a very efficient way of controlling the tension and compression forces within it. Imagine that an axial force is being applied to the concrete—as though we were trying to squeeze it—so that a concrete beam, for example, is put it into precompression before any load is applied. Well that is what is being done in prestressed concrete.

So why is it so useful? By putting the concrete into precompression it will not go into tension when a load is applied, because the precompression stress in the concrete can be made larger than the tension stress from the bending. This is a very simple explanation of how prestressed concrete works. It is much more complex than that in the design of a bridge beam because there are losses in the prestressing strand due to shrinkage of the concrete, and relaxation of the steel stress.

When strands are placed in beams or box girders and tensioned before the concrete is placed in the forms, it is called prestressing. When the strands are tensioned after the concrete has hardened, it is called post-tensioning. In prestressing, the tension wires or strands are cut once the concrete surrounding them has hardened, thus transferring the force from the strand to the concrete. In post-tensioning, the strands are laid in position but are not tensioned.

When the concrete has hardened, anchor plates are placed at the ends of the beam and the strands are tensioned against them, putting the concrete into compression. With prestressed concrete the depth of the beam or box girders is usually 20 percent shallower than the equivalent reinforced-concrete beam, because prestressing is more efficient. It is the preferred method of construction for box-girder beams and bridge decks in modern concrete bridges.

BELOW LEFT: Explaining reinforced concrete.

BELOW RIGHT: How prestressed concrete works.

Plain concrete

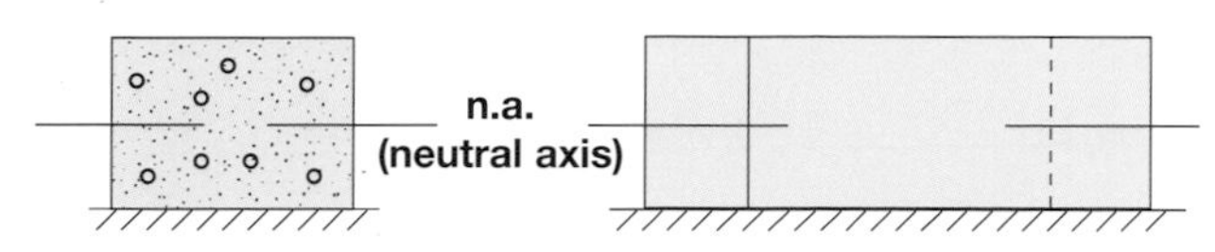

Plain concrete beam resting on the ground has zero stress.

zero stress

Reinforced concrete

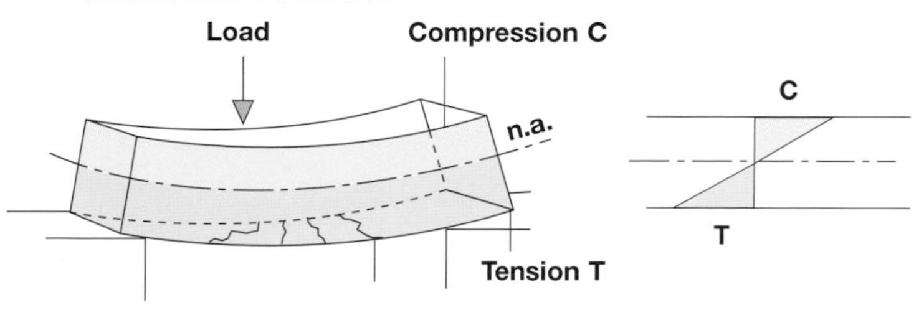

If a plain concrete beam has to span it will bend—the top half above the neutral axis or the line of zero stress will be compressed, while the bottom half will be stretched and will go into tension.

Bending

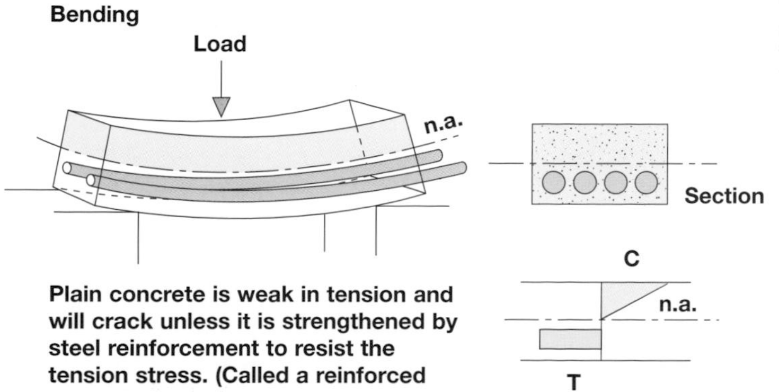

Plain concrete is weak in tension and will crack unless it is strengthened by steel reinforcement to resist the tension stress. (Called a reinforced concrete beam.)

Prestressed concrete

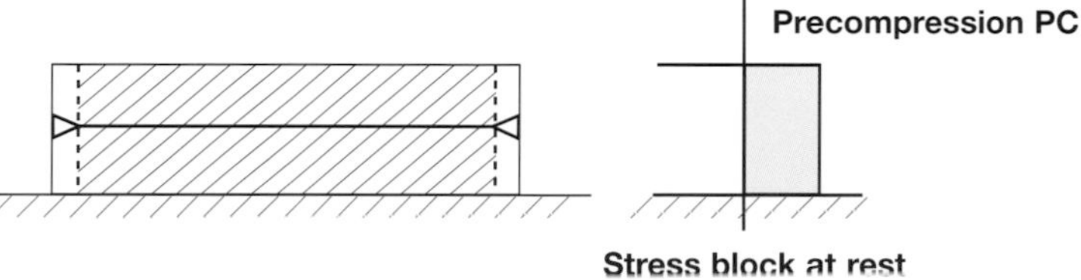

A plain concrete beam is prestressed by tensioning steel wire strands that run inside the beam, and which are anchored at each end of the beam. The steel strands compress the beam and impart a pre-compression.

Load

PC **C** **C** **n.a.** **T** **C**

at rest **load** **final**

When a load is now applied, the prestressed beam does not go into tension when it bends, because the pre-compression stress imparted by the steel strands is much greater than the tension stress. A prestressed beam is more efficient than a reinforced concrete beam.

The steel beam

Steel is processed and manufactured in factories and then assembled into beam sections, plate girders, trusses, and so forth at steel-fabrication yards, before they are transported to the bridge location. Steel is stronger than concrete and has an elastic modulus that is 15 times greater, making it 15 times more efficient than concrete, but also about that many times more expensive.

To determine the most efficient section and shape for a steel beam, we must consider how it behaves under bending. In bending, the lower part of the beam stretches and goes into tension while the upper part shortens and goes into compression, and in the middle, or neutral, axis it neither stretches nor shortens. This means that the steel at and near the neutral axis does very little work to resist bending, while the steel farthest away from the neutral axis has to resist most of the load. Because of this it is more effective to remove as much of the steel as possible near the neutral axis and place it at the ends. If we start with a beam with a solid rectangular section and then take away the inefficient material near the neutral axis and put it symmetrically about both ends, we get a wide-flange section or I-beam. This is the most efficient shape for a steel beam and, since bending resistance depends on the distance from the neutral axis, the deeper the beam, the greater is the bending resistance of a wide-flange section. The central section of the beam is called the web.

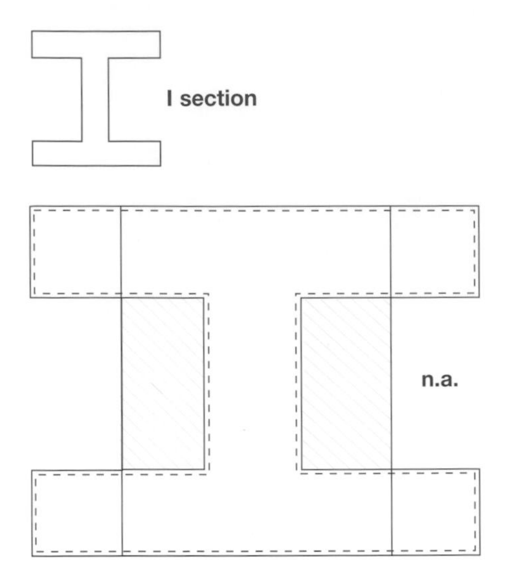

ABOVE: A steam beam section

If we take the inefficient material away from near the neutral axis of a square section beam and place it symmetrically at either end, we obtain an efficient I section steel beam.

BELOW: Example of steel I section beam for a bridge span (Viaduc du Pau, France).

The arch

Early arch bridges The arch bridge is a pure compression structure, and such bridges were in common use at the time of the Roman Empire. Because only brittle materials such as stones and bricks were available, to span any distance more than a large hop they had to employ arch construction to ensure that the forces carried by these materials were always kept in compression.

The Romans built their arches in the shape of a semicircle. It was a simple geometric shape, making it easy to form the centering that supported the wedge-shaped stones, or voussoirs, for the arch. The voussoirs were cut very accurately to the maintain the circular profile of the arch. They were placed symmetrically on the centering, working from both ends of the arch toward the crown.

When the center voussoir, or the keystone, was wedged firmly into place, the centering was taken down. Basically, an arch consists of two halves that lean against each other at the keystone. The weight of the bridge is carried outward along the curving path of the voussoirs. At the

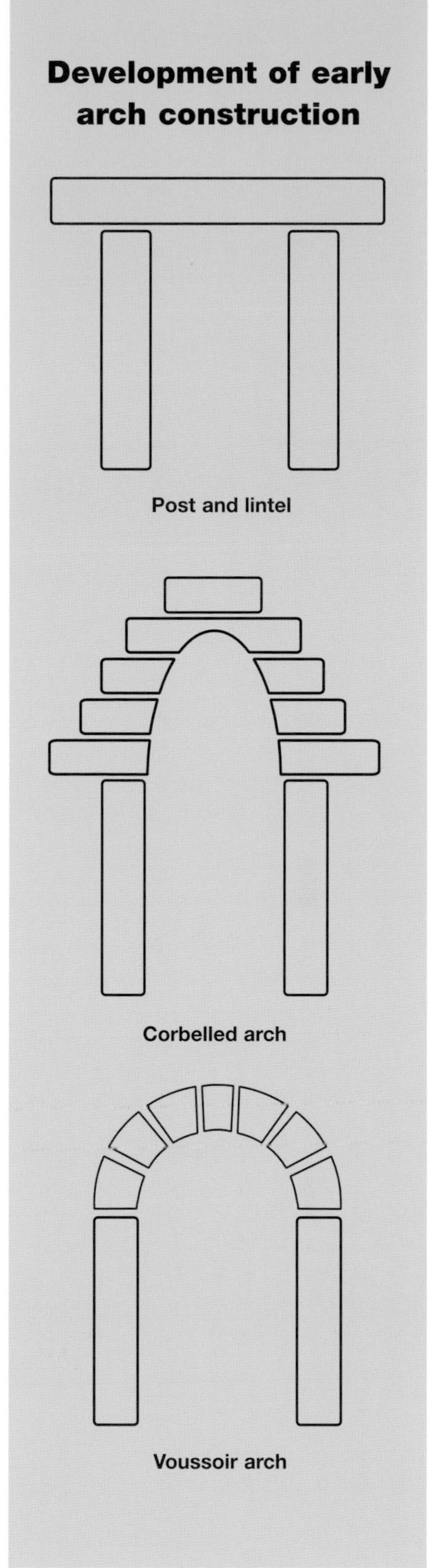

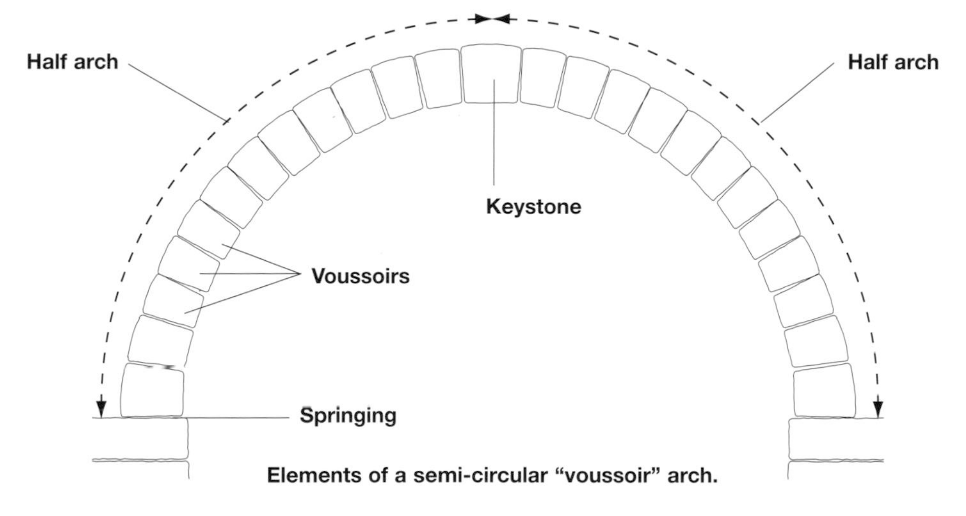

Elements of a semi-circular "voussoir" arch.

TOP LEFT: Roman arch construction (Segovia Aqueduct, Spain).

LEFT: Peunta Alcantara over the Tagus in Spain (98 AD).

ABOVE: Example of an elliptical arch. The circular apertures in the piers are for flood waters to pass.

point where the arch reaches the ground, the downward force from the arch is resisted by piers, while the outward thrust of the arch is resisted by the abutments to prevent it from spreading out and thus collapsing.

For a semicircular arch such as a Roman arch, which met the ground vertically, there is no outward thrust. Thus each arch span can be built independently of the next one, which was useful in Roman times, when they had to build a bridge across a river. However, the pier supports for such arches have to be very wide and massive, and the span quite limited. The flatter the curve of the arch, the greater is the outward thrust on the abutment, and the more unstable the arch becomes during construction.

The two-centered, or pointed, arch is one in which the two halves were formed from the arcs of a circle that meet at a point in the middle. These were built to provide better navigation clearance under the span.

There are different types of arch bridges. The reason for this is that some are more efficient at spanning greater lengths, while others are more economical in the use of materials, and yet others may be faster to assemble.

The segmental arch The segmental arch has a profoundly different structural effect from that of the semicircular arch. The segmental-arch profile is part of an arc of a larger circle. As a result of this the segmental arch introduces sideways or horizontal thrust, because it does not press down wholly in the vertical direction when it meets the ground. The flatter the arch the greater the horizontal thrust.

The advantage of the segmental arch is a flatter arch profile that spans farther and requires less material in constructing the arch and roadway it supports. Requiring fewer piers in the stream or river crossing meant less obstruction to navigation and to seasonal floodwaters. The abutments built at each end of the bridge receive the horizontal thrust from the segmental bridge span.

In a bridge with more than a single span, the horizontal thrusts of adjoining spans are equal and opposite. Thus multiple arches could be built to hold one another up, as long as they were supported by abutments at each end of the bridge. Earlier bridges needed heavy piers to act as abutments in order to build the spans independently. This meant that the piers could be made slimmer because they had to carry only the vertical load, since the arch spans were interdependent.

The stiffened-slab arch This was a discovery of Robert Maillart, who exploited the plastic qualities of reinforced concrete to build very slender, remarkably modern-looking, three-hinged and stiffened-slab-arch bridges at the turn of the century. At each abutment the arch begins as a pair of legs which thicken and fuse toward the quarter-span. At this point the arch joins the bridge deck or road slab.

Over the central section the bridge deck and arch come together to form a stiff box section. The forces in the structure are designed to concentrate at the three points or hinges—one at each abutment and one at the crown. The stiffened-slab arch was characterized by a rigid and heavy bridge deck supported on a thin slender arch slab via pencil-thin, vertical walls. The rigidity of the deck restricts any lateral movement of the bridge because it is joined to the arch at the crown and anchored firmly at each end. It also distributes the load from the bridge deck evenly over the whole bridge, so that buckling in the arch is prevented.

TOP: Example of Maillart's stiffened-slab concrete arch. Schwandbach Bridge, Switzerland.

ABOVE: Tavansa, Switzerland, another stiffened-slab arch bridge designed by Maillart.

Modern arch bridges Modern arch bridges are built mostly of reinforced concrete and steel. The concrete of the arch is poured into wooden or metal forms supported on a scaffolding, once the steel reinforcement is in place. The formwork is removed when the concrete has matured and fully hardened. Usually a concrete-arch bridge consists of two or more parallel arches, from which columns of varying lengths rise to support the bridge deck. The parallel arches are interconnected by struts crisscrossing each other, which makes the arches work together against the lateral pressure of the wind.

Steel-arch bridges are built very similarly to concrete-arch bridges, with a series of wide-flange plate-girder or box-girder beams or tubular-steel elements, which are prefabricated in sections, to the required profile of the arch. These

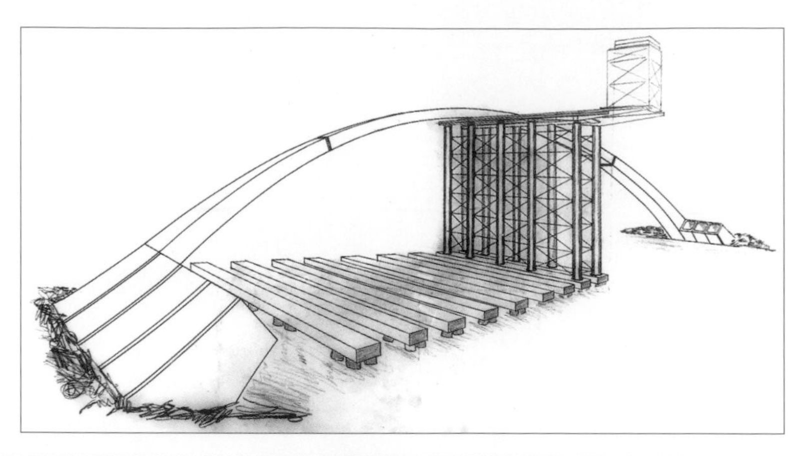

elements are transported to site, and temporarily supported in position, until they are welded and bolted together. Stub columns of steel of varying lengths are built up off the arch to support the bridge deck platform. Steel arches have certain advantages over concrete arches, since they are lighter in dead weight, require smaller foundations and need less supporting scaffolding. They can also be built by cantilevering each half of the arch and connecting them at mid-span.

Modern arch bridges can form slender, flat-arch profiles or very pronounced segmental-arch curves, depending on the length of the span required and the ground conditions. If the ground is very poor on the embankment for instance, but better under the river bed, it may be economic to build a deeply curved segmental arch with less horizontal thrust than a flat-arch bridge.

Variations on the arch shape have been developed to exploit its lightweight construction, span range, navigational clearance, and ability to increase or reduce the horizontal thrust depending on the ground conditions. There is the through arch, sometimes called a "sickle arch," where part of the bridge deck is suspended from the arch and part of it supported by columns from below.

TOP: Illustration showing the scaffold support for the concrete arch span of the Gladesville Bridge.

ABOVE: Gladesville Bridge in Sydney under construction.

RIGHT: Example of a steel flat arch (Pont St Ouen, Paris, France).

ABOVE: Gladesville Bridge, Sydney, Australia.

LEFT: Types of modern steel and concrete arch bridges.

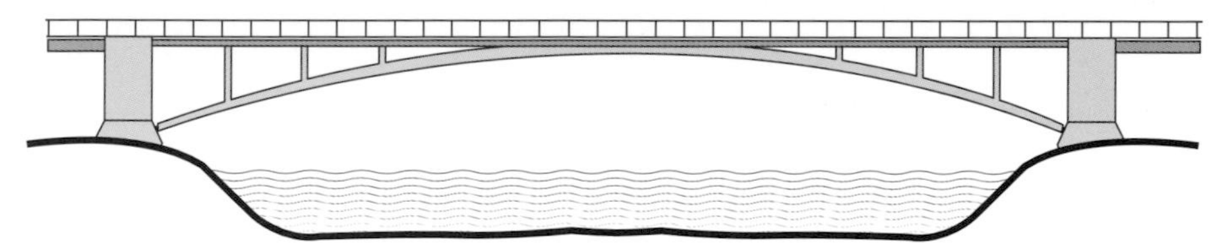

The two hinged arch

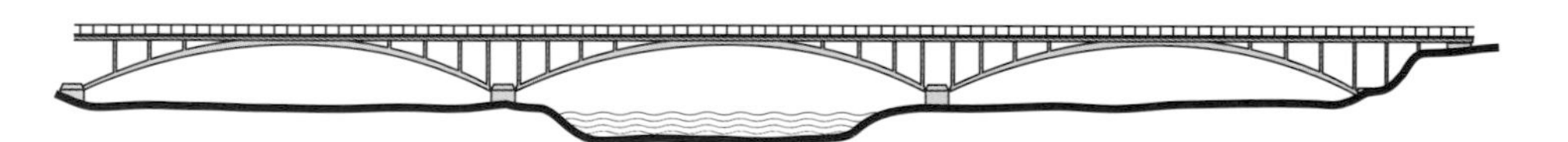

Sequence of flat two hinged arches

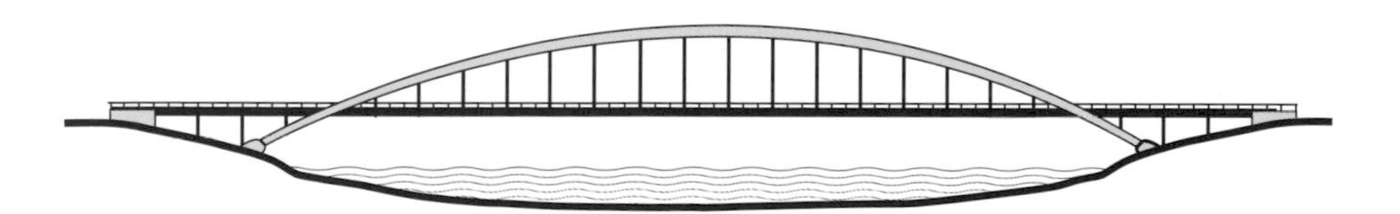

Sickle-shaped through arch with suspended deck

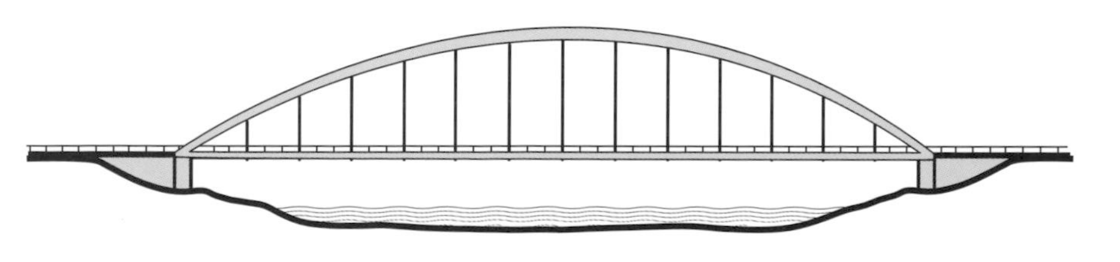

Bowstring, tied arch

ABOVE: A concrete tied arch bridge under construction (El Rincon Viaduct, Spain).

ABOVE: Constructing the steel arch of the Henry Hudson Bridge, New York.

There is also the "bowstring arch," or tied arch, where the bridge deck is suspended from hangers or truss bracing from the arch. The arch is tied by the span of the deck, which resists the outward thrust of the arch, so that no abutments are required. This type of arch was popular for rail and road bridges across the world, where the river bed was deep or the embankment was quite shallow, making navigation clearance critical.

RIGHT: The completed steel arch of the Henry Hudson Bridge.

Truss girders

One of the most simple and basic of bridge forms is the triangular truss. A simple triangular truss is made up of two inclined compression members and a horizontal tension bar or tie rod, which prevents the inclined members from opening up under load. The truss is self-contained and is supported from each end, to carry the imposed load as well as its own weight. It does not need abutments because there is no horizontal thrust.

In some long-span trusses, the sagging of the tie rod under its own weight is reduced by support at mid-span from a hanger suspended from the top of the truss. Large trusses or composite trusses can be made by connecting triangular trusses together. Highway and railroad bridges spanning hundreds of feet are often built with trusses like this, mostly made of steel bar.

The composite truss behaves very like a deep-flanged beam, with the truss upper chord acting like the upper flange of a steel beam, and the lower chord like the lower flange of a beam, and the inclined or diagonal bracing bars as the web

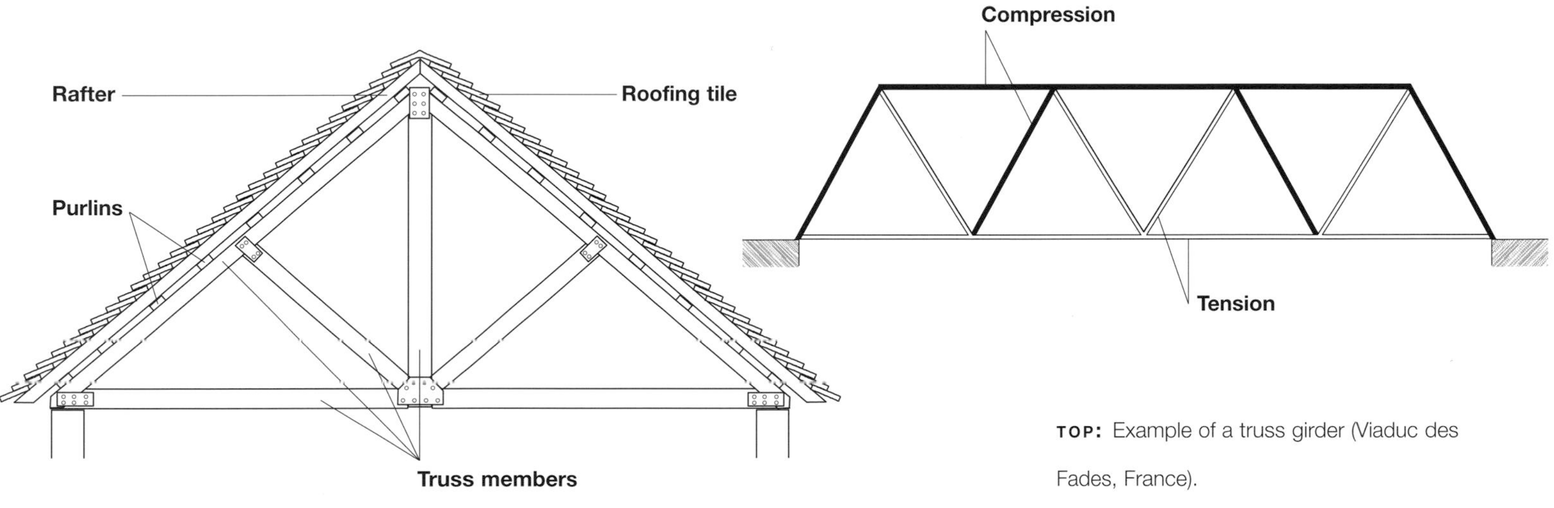

TOP: Example of a truss girder (Viaduc des Fades, France).

ABOVE LEFT: Simple timber roof truss

ABOVE RIGHT: Long span steel truss.

LEFT: Walnut Street Bridge, USA (1889).

ABOVE LEFT: Central truss span of the Carquinez Bridge being erected.

ABOVE RIGHT: Carquinez Bridge, USA (1927).

of the steel beam. The diagonals of the truss connecting the upper and lower chord of the truss are much lighter than a full web and use less material. The rigidity and stiffness of the truss is due to the triangulated diagonals.

Gustave Eiffel's great bridges and towers were based on the lightness and rigidity of the truss, because it was very strong and yet economical on materials.

Box-girder and trapezoidal box bridges

Steel beams, plate-girder beams, and reinforced- and prestressed-concrete beams are suitable for relatively short spans for road and railroad bridges, but will require many supporting piers if they are to be built across a wide river or freeway interchange. By forming a box section which is hollow in the middle, a very strong but economical girder beam can be built, which has a greater span range. The box-girder beam, whether it is made of steel or concrete, is very common for modern road or railroad viaducts, and has been used extensively for building elevated freeways along mountain ranges or over valleys in Europe. It was primarily developed for the rigid deck construction required for suspension or cable-stay bridges. The box-girder bridge was first conceived by Robert Stephenson for his Britannia Rail Bridge over the Menai Straits in Wales.

BELOW (TOP TO BOTTOM):

Stee' plate girder.

Steel box girder, with inclined struts.

Concrete box girder.

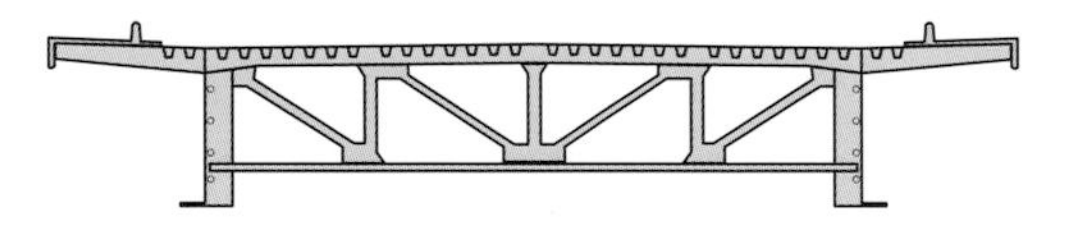

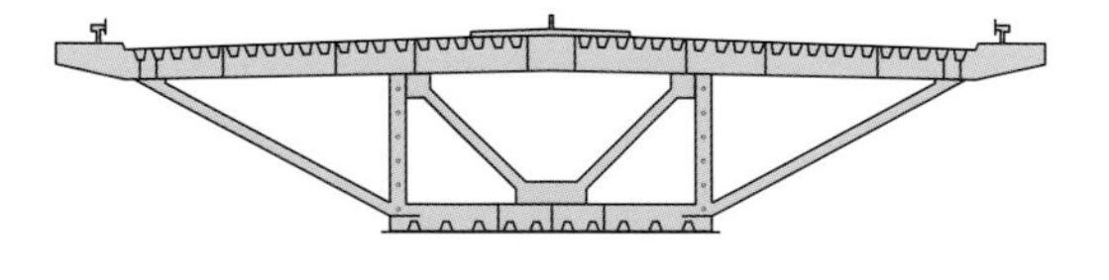

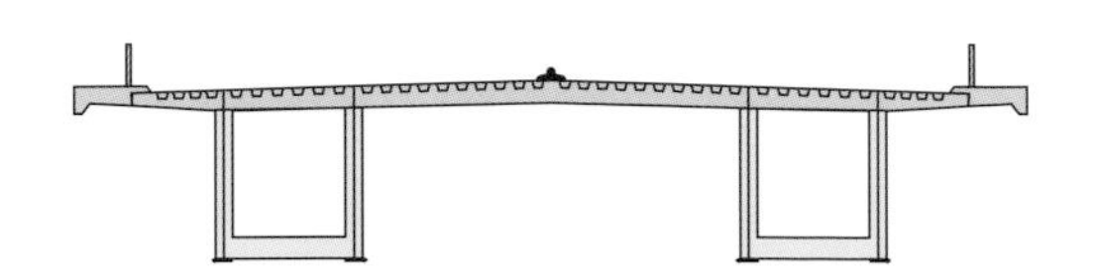

RIGHT: Crane erected, balanced cantilever, cast-in-place box girder construction.

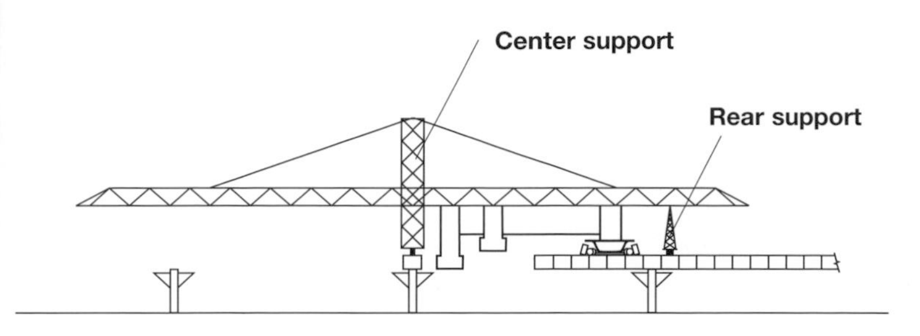

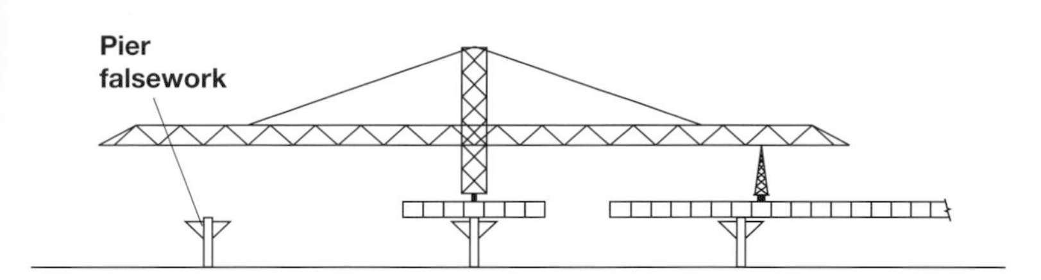

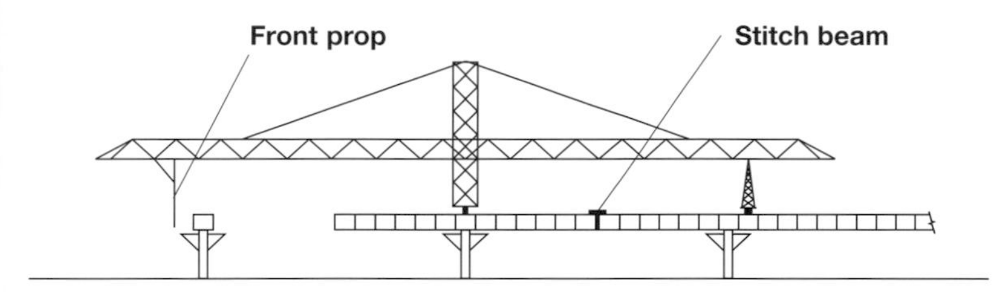

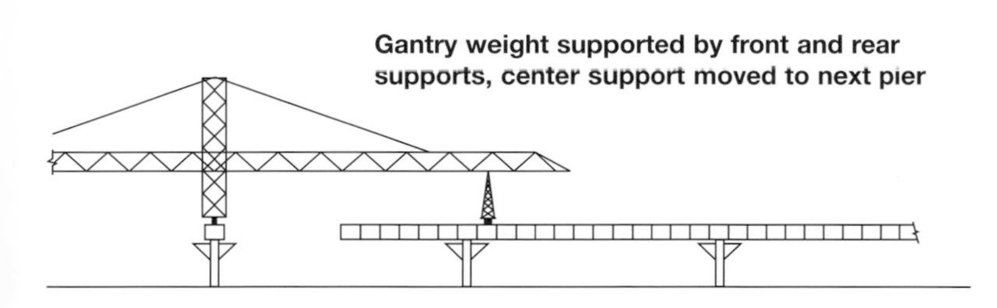

ABOVE: Sequence of box girder cantilever construction.

TOP: Example of a steel box girder with inclined struts.

ABOVE: Cast in place, cantilever box girder construction.

LEFT: Gantry erected precast box girder span.

OPPOSITE: The suspension cable and walkway of the Akashi Kaikyo Bridge, Japan.

OPPOSITE RIGHT: The deck structure is suspended from hangers connected to the main cable (Kurushima Bridge).

FOLLOWING FOUR PAGE SPREAD: "Record Breaking Suspension Span" of the East Bridge of the Great Storebelt Crossing, Denmark.

Suspension bridges

Natural vegetable fibers like hemp and bamboo have been used for thousands of years to make strong ropes from which to suspend crude bridges or to secure masts and sails in strong winds. Nowadays we make thick steel rope by twisting individual steel wires to make a cable strand, which can measure a few inches to a few feet in diameter. The steel cable is protected by a plastic nylon sheath to prevent corrosion. Such cables can carry thousands of tons of force and are the basic load-carrying system of the longest type of bridge in the world—the suspension bridge.

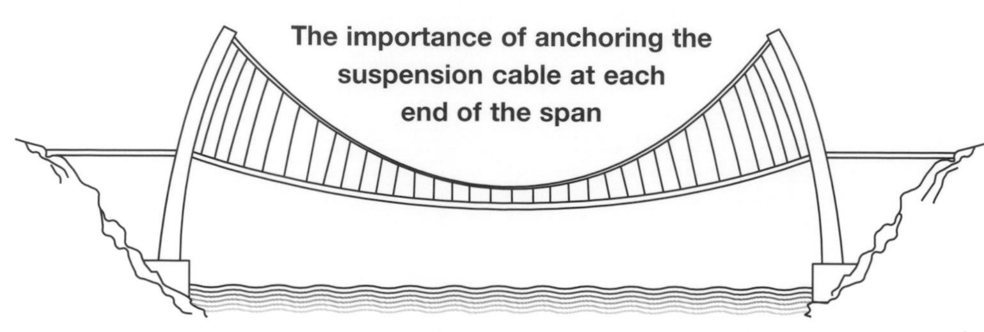

With no end anchorages: the suspension cable transfers the vertical load and the tension in the cable to the towers, which become unstable and bend critically inwards.

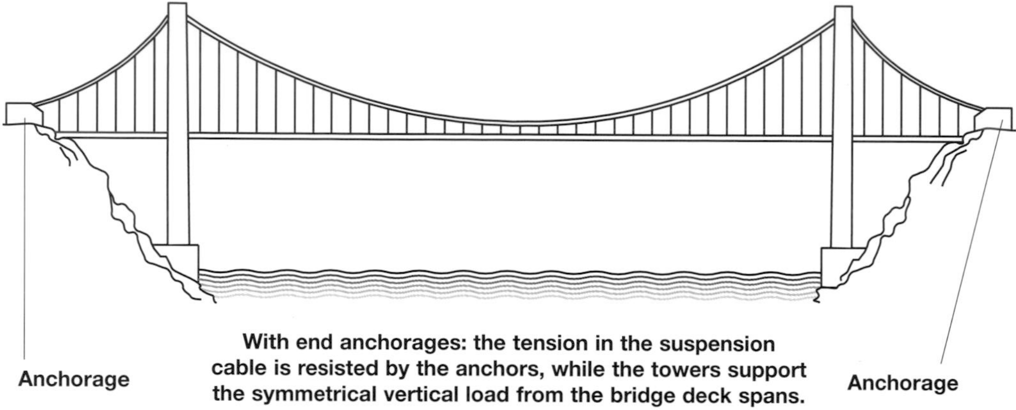

With end anchorages: the tension in the suspension cable is resisted by the anchors, while the towers support the symmetrical vertical load from the bridge deck spans.

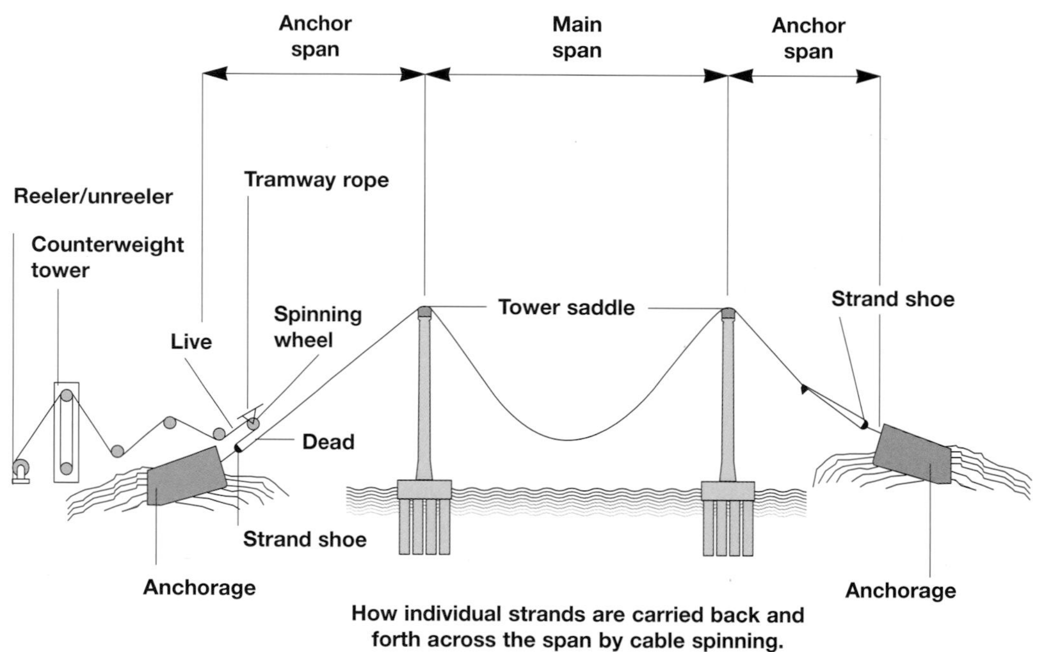

How individual strands are carried back and forth across the span by cable spinning.

A cable is a tension structure because it can work only in tension. But you cannot build a bridge with just a series of cable strands, since the cable must be supported by towers and anchored at each end to maintain its suspension or catenary shape. So here we have the basics of the suspension bridge: the tension cable and hangers; the support towers and foundation, or caissons, which are in compression; the roadway, which is a truss or stiff box girder providing the lateral stiffness against wind pressure; and the end anchors, which stop the towers bending and resist the tension pull in the cable.

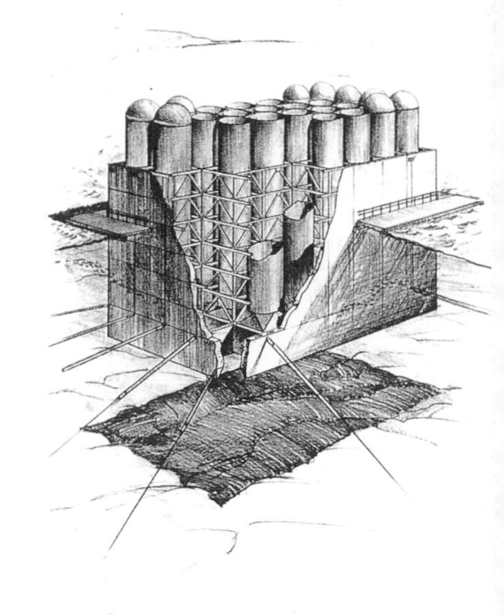

RIGHT: Illustration showing how the caisson foundations and piles for the piers of the Tagus Bridge, Portugal, were constructed.

LEFT: Hanger connections being checked from the main suspension cable.

BELOW: Spinning the individual strands of the suspension cable.

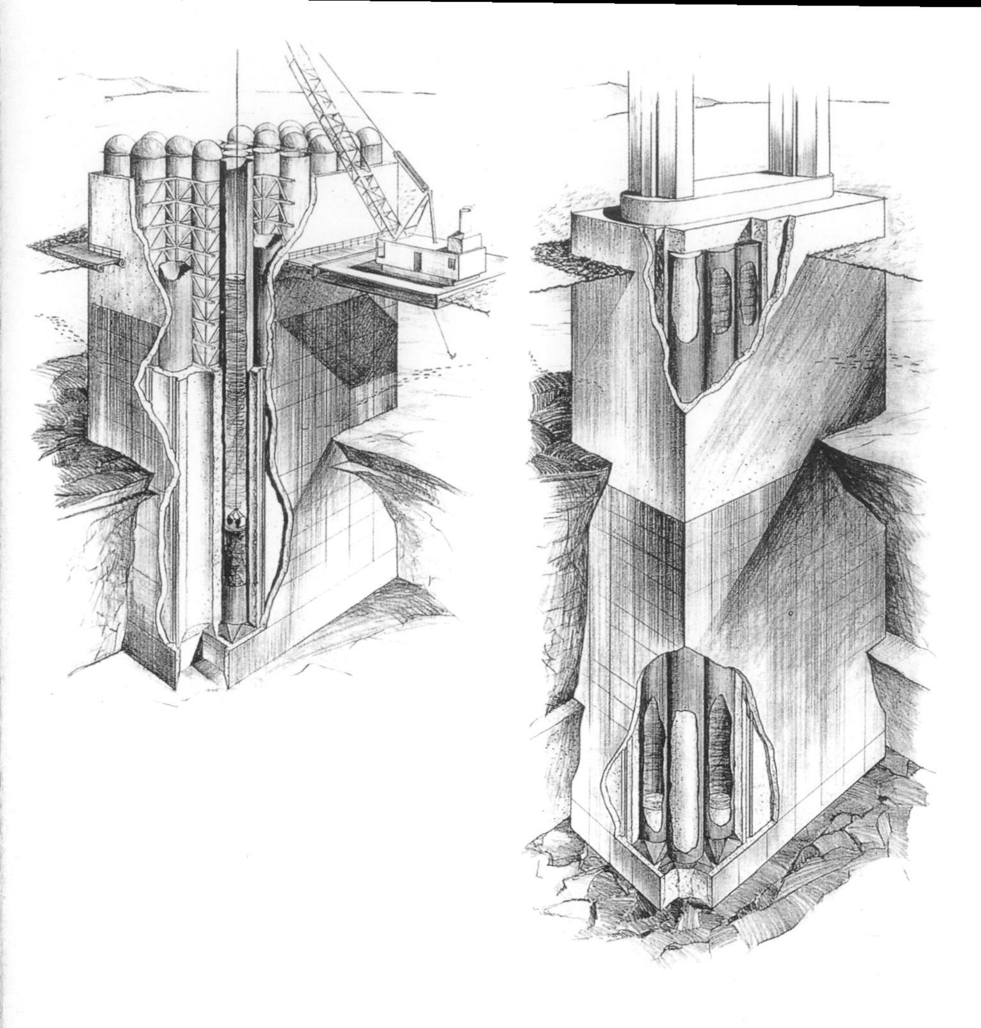

How does the suspension bridge keep its shape as traffic moves over it? The cables, which will want to distort under load, are stabilized by the hangers that support the roadway from the suspension cable. As a load goes over the bridge at a particular point, the cable is held in shape by the hangers and the rigid bridge deck. To keep the horizontal distance between the cables constant, the bridge deck acts like a stiffening truss and maintains the shape of the suspension system.

The larger the sag of the cable between the towers, the less tension is in the cable, which in turn means smaller-diameter and less costly cables and smaller end anchors. The problem is that, in creating a larger sag, the towers that support

ABOVE: Looking through the cable saddle at the top of the tower of the Mackinac Bridge.

LEFT: Floating out the caisson and pile tubes for the Tagus bridge foundations.

ABOVE: Spinning the suspension cable on the Mackinac Bridge.

the cables have to be built very high, and this extra cost often outweighs the saving in the cable costs. That is why many suspension bridges have a shallow curvature of their suspension cables. However, the shallow curvature increases the tension in the cables, so thicker cables and large anchor blocks have to be built to resist this force.

The cables of a suspension bridge can weigh hundreds of tons, making it impractical to make the entire cable on dry land. Instead the cables are "air-spun," very like in a textile loom, with individual steel wires pulled across in pairs by a spinning wheel. To set up the spinning wheels, pilot ropes are taken across the span and hauled up over both towers and linked back to the anchors at each end. A catwalk is then assembled from the pilot ropes for each cable, suspended a few feet below the eventual position of the cables.

The spinning guide wires are then sent across and large pulley wheels—one at each end—create a continuous loop for the spinning wheel to track. A section of steel wire is then attached to each of the two spinning wheels which pulls the wires across the entire span. Each time the spinning wheel covers the entire span

it lays two wires fed from two large reels of wire. The spinning wheel runs the continuos loop back and forth across the span until the required number of wires have been positioned.

When the spinning wheel runs over the tower, the wires are placed in special metal saddles at the top, which are carefully located to ensure there is an even spread of the compression force pushing down on the tower from the tension in the cable. The wires are packed tightly together by a special clamping machine, then covered with wrapping wire, painted, and sheathed in a plastic nylon covering for protection. Cable bands are clamped at regular intervals, to attach the hangers that support the bridge deck.

ABOVE: With the anchor blocks and the suspension cable in place, the approach span construction begins on the Mackinac Bridge.

BELOW: The cable strand shoes of the anchor block of the mighty Verazzano Narrows Bridge.

Usually suspension bridges have caisson foundations, which for the East Bridge of the Great Storebelt crossing in Denmark were prefabricated in concrete in a dry dock from where they were floated and then towed out to the center of the estuary. The caissons were ballasted for buoyancy during towing and, once in position, they were flooded to sink down onto the prepared bed on the bottom of the river estuary. The caissons were injected with sand and concrete, displacing all the entrapped water within their voided compartments to form a rigid and stable foundation. The suspension towers are built up from the caisson.

RIGHT: Segments of the truss deck structure are attached to the hangers of the Verazzano Narrows suspension bridge.

BELOW: The East Bridge of the Great Belt crossing in Denmark.

Cable-stay bridges

This type of bridge is a relative newcomer in the bridge world and was developed during and immediately after World War II. They are more efficient than suspension bridges over shorter spans, requiring fewer cables to support the bridge deck. Instead of long draping suspension cables, the bridge deck is supported by a series of individual cables connecting the bridge deck directly to the pylon or cable mast, with the cables arranged in either a fan configuration or a harp configuration. In the fan configuration all the cables run over the top of the mast via a saddle, while in the harp configuration they run parallel to one another and are connected at equal intervals along the bridge deck and the cable mast.

All the early cable-stay bridges used two planes of stays, with cables from twin towers supporting both edges of the bridge deck. The Severins Bridge at Cologne was the first to use a giant A-frame tower to support the fan stays of this asymmetrical-span bridge.

As the trapezoidal box-girder bridge deck was developed for the suspension bridges, to provide an aerodynamic shape to minimize buffeting effects of the

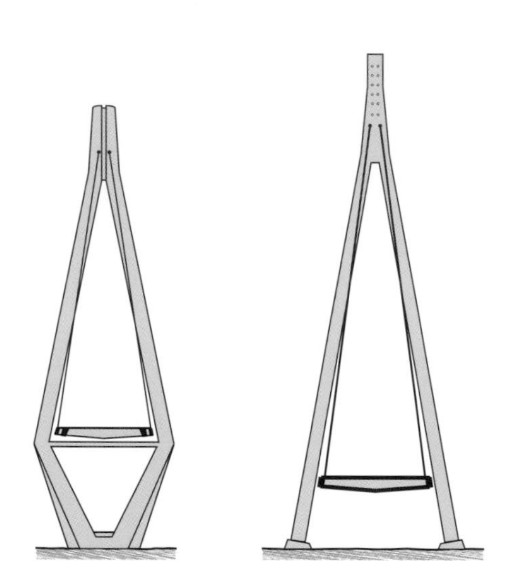

ABOVE: Shape of pylon towers for supporting a "double" cable stay arrangement.

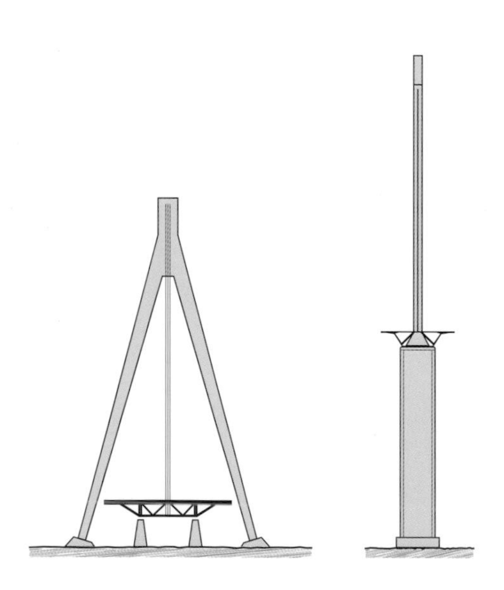

ABOVE: Shape of pylon towers for supporting a "single" cable stay arrangement.

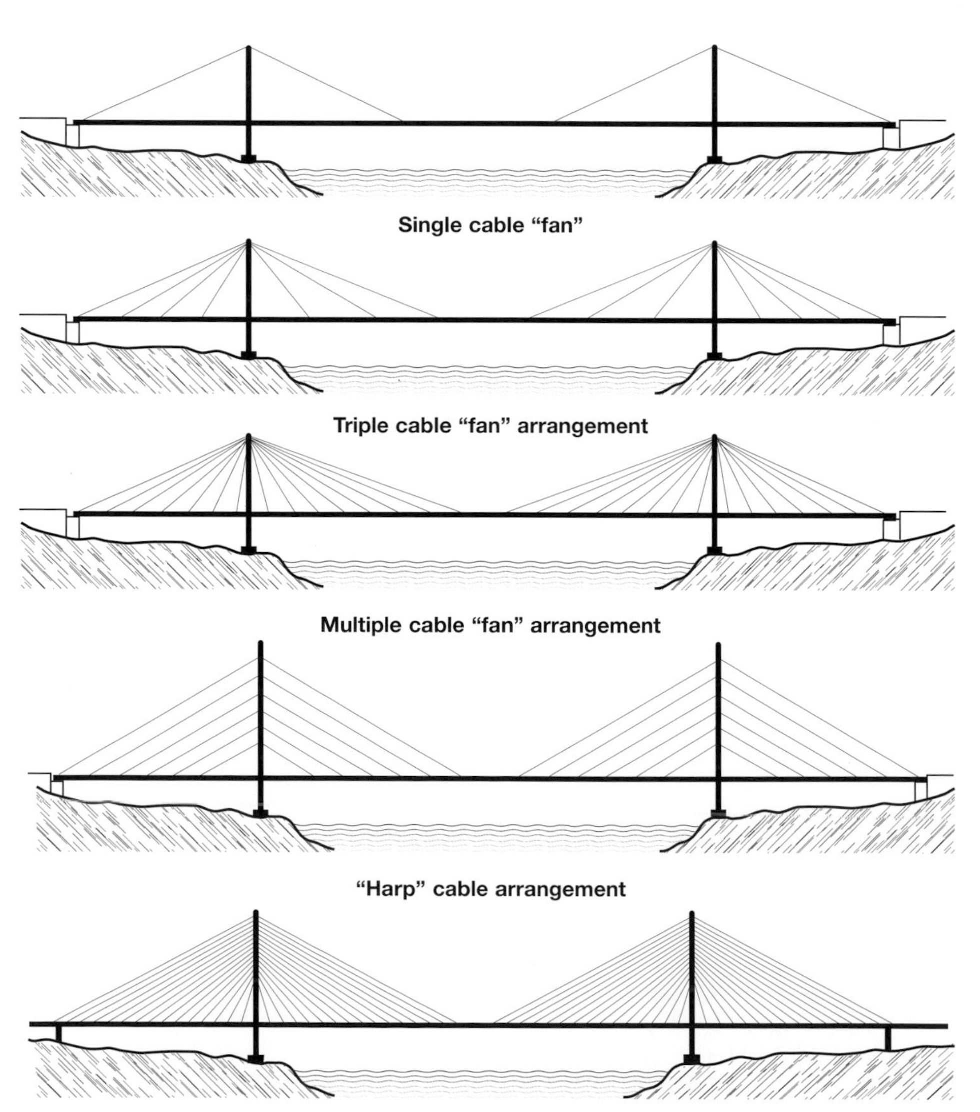

RIGHT: Different types of cable stay bridges.

TOP: The pile foundation supports for the approach spans.

ABOVE: Erecting the concrete box sections by crane for the approach spans.

TOP: Precast concrete segments in the casting yard.

ABOVE: Completing the box girder approach span.

ABOVE: Constructing the pylon superstructure for the cable stay span.

wind, they became torsionally stronger and more rigid. This allowed cable stays to support the bridge deck with just a single plane of stays, instead of a pair for each side of the deck. The stays are connected along points within the central reservation of the road or bridge deck. Visually the single plane of stays is more elegant and less fussy than two planes of stays.

For a cable-stay bridge to work properly the cables themselves have to be extremely powerful and must act on the bridge deck with as much rigidity as possible. In the early days several bridges were not successful, since the cables "stretched" too much under cyclic loading. To counter this the modern stay cable is "lock-coiled," whereby the strands in the cable are specially wound to restrict their unwinding under tension.

Constructing the Vasco da Gama Bridge, Portugal This is an immense bridge project which took many years to plan and design and just 18 months to construct. It represents a world record for a bridge of this scale. The central navigation span is a cable-stay bridge, while the others, called the approach spans, are shorter, concrete, box-girder spans, which have been built using the balanced cantilever method.

Here's a short summary of the construction stages to highlight the main features of the project.

1. Prior to starting any bridge work a marshaling yard, a precast production yard and fabrication site is established. This is where all the materials and equipment for building the bridge are delivered or made—for example, the steel sections, the erection gantry, the precast-concrete box girders, the huge caisson foundations, the scaffolding and temporary works, the cable-strand reels, the cranes, the piling rig, the access platform, the floating pontoons, the dredgers, and so forth. The site can be as large as 74 acres, with offices and parking lots to accommodate over 3,500 people.

   In the case of the Vasco da Gama Bridge, the foundations for the central span's piers were the first to be completed. They consist of 44 bored piles which were 165 feet long and just over 7 feet in diameter, surmounted by the pylon base. The base is a concrete raft 275 feet by 65 feet in plan and about 10 foot thick, and has prestressed beams, which connect the two legs of the H-shaped pylon to the base. The pylon base ensures the distribution of the vertical forces from the bridge and resists the horizontal thrust which may arise owing to impact from shipping.
2. The pylon legs were then built up from the base, by a climbing formwork system. The reinforcement for the pylon legs was placed in the 13-foot-high formwork section, before the forms were clamped. Concrete was poured into place, and the forms were eased away once the concrete had hardened. The forms were crane-lifted to the next 13-foot-high section, and adjusted to the

correct width and taper of the pylon profile, before being concreted again. Stage by stage, the climbing formwork rose to create the legs of the 490-foot-high shafts of the twin pylon supports.

3. Next the pylon crossbeam was cast, before the cable support section of the twin pylon masts was built. The top section of the pylons was formed using precast-concrete box elements that stacked one on top of another, all the way to the top. The box sections were precast on shore and transported by barge to the bridge site. Each section was lifted onto the previous section by a barge-mounted crane, and then jointed using *in-situ* concrete. Steel sections were cast into box sections to provide the anchorage for the cable stays.
4. The prefabricated steel formwork to cast the bridge deck was then assembled at the base of each pier. The pier section of bridge girder deck was cast and the 2,000-ton assembly was then jacked into position. As the bridge is in a

BELOW LEFT: Building out the central span.

BELOW RIGHT: Connecting the cable stays to the bridge deck.

ABOVE: "The gap to span"—building the bridge deck of the main span.

BELOW: Cable covers protect each of the cable stays.

seismic zone and has to withstand earthquake movements of up to 4 feet 3 inches across the width of the span and 6 feet along it. In order to withstand this movement the bridge deck is suspended directly from the pylon and not connected to the crossbeam on the pylon masts, which is more usual. Damping devices diminish the deck movement and ensure the bridge-deck-to-pylon linkage.

5. Once the pier section of the bridge deck was in position, the steel formwork gantry assembly was cantilevered out from both ends of the pier section, to allow 29-foot sections of the bridge deck to be concreted. The deck assembly was held in position by the permanent cable stays, which were connected to the pylon mast. Slowly, in 29-foot lengths, the bridge deck started to span across the 1,380-foot central gap.

   Built as an extension of the main span, the central viaduct has a length of 21,370 feet and is composed of 81 bays in total, 75 of which span 258 feet, two span 426 feet over a navigation channel, and four span 308 feet. In cross-section the concrete box-girder deck consists of two parallel girders, one for each carriageway, and has a standard depth of 13 feet, which increases to 26 feet for the longer spans, with the top section of the box girder forming the roadway slab. The box girders are haunched (get deeper) over the pier supports, creating a flat-arch profile.

6. The concrete box sections were cast 15 miles upstream of the bridge. Each section was between 23 and 33 feet long, weighing around 256 tons. Cast on an assembly line, the box sections were connected in groups of eight to form a prestressed beam 256 feet long and weighing 2,165 tons. They were towed out in pairs and lifted into place by barge-mounted cranes, and then stitched onto the central pier section using prestress, to maintain the balanced cantilevers. The southern approach viaduct has an overall length of 12,550 feet and was built using gantry-launched, balanced cantilever construction. The four deep beams for each span were cast in place, using movable formwork.

On March 31, 1998 the Vasco Da Gama Bridge was officially opened.

TOP: Nearly two-thirds of the cantilevered half span is now complete.

ABOVE: The view looking along the centerline of the finished cable stay spans.

LEFT: Aerial view of the distinctive cable stay spans of the Vasco da Gama Bridge over the Tagus estuary in Lisbon, Portugal (1998).

4

Great bridges—the bridges of destiny

It is never easy to make a definitive list of the greatest bridges of the world. In making a short list, it seems logical to select bridges that are representative of the great periods of bridge building over the centuries, such as the Roman, Medieval, Renaissance, the Age of Iron, the Industrial Age, and so on.

At the same time we are also drawn toward a number of spectacular bridges that were built during this century; but these have been depicted in every book on bridges in the past 20 years.

Taking my cue from the great bridge-building eras, I have made my selection with the proviso that the bridge must still be standing today, it must be accessible to the public, and must sit in a wonderful location. It was also important that the images available should be stunning and, for the contemporary subjects, images of the bridge taken both during construction and after completion should be shown, to contrast that with pictures taken more recently, if that was possible. It is important to highlight the working conditions during construction—the dangers and the risks—and the plant and equipment that were used to build the bridge. It gives a sense of the scale and hardship that were endured during construction, which may never be fully appreciated when viewing the finished bridge.

ABOVE: Building the cantilever steel arch of the St Louis bridge over the Mississippi, USA.

OPPOSITE: The Sunshine Skyway Bridge in Florida hit by lightening during a thunderstorm.

The bridges I've chosen are:

Pons Augustus
Valentré
Santa Trinità
Rialto Bridge
Pont de la Concorde
St Louis
Garabit Viaduct
Firth of Forth
Plougastel
George Washington
Sydney Harbour
Golden Gate
Severn
Sunshine Skyway
Pont de Normandie

Pons Augustus, Rimini, Italy (*c.* AD 14) Many consider this bridge over the Marecchia at Rimini the finest bridge of the ancient world and greatest bridge ever built by the Romans. It was built during the reign of Augustus (31 BC—AD 14). The genius of Roman bridge engineers seen in the Pons Augustus, the Puenta Alcantara and Segovia Aqueduct in Spain, Trajan's Bridge over the Danube, and the Pont du Gard in Nîmes expresses the power and might of the Roman Empire better than all the ruins of the Forum in Rome. Palladio in the sixteenth century declared Pons Augustus the finest bridge in the ancient world and copied the spans many times in his own construction of bridges. And, because Palladio's bridge designs were much admired, they were also copied by bridge builders in later centuries. We can see second- and third-generation "Pons Augustus" in many European cities today.

The Pons Augustus has five spans, the three middle ones are 28 feet, and the two end ones are 23 feet. The whole structure is unifying. On each spandrel above the piers, there is a panel framed by pilasters that uphold a classic pediment. The balustrade is solid, its heavy cornice supported on toothlike stone supports called dentils, while the face of the bridge was covered in marble. It was unusual for a bridge built outside Rome to feature such charming decoration.

From a structural viewpoint the Pons Augustus is interesting because it is the earliest known example of a bridge built on a skew. The piers are not at right angles with the axis of the bridge, although the amount of skew is small. It is probable that the bridge builders did this to locate the pier foundation in the river bed, where the soil and the current made construction easier.

When the construction methods of the Romans are recalled, the Pons Augustus, along with all the other great bridges, will be remembered with awe. Generally, slaves did the laboring, hauling the stones, lashing and nailing the timber centering, building cofferdams in fast-flowing streams, under Roman supervision. Crushing and drowning must have been frequent. They used the simplest of tools—the wedge, the lever, the pulley, and the inclined plane. For Trajan's Bridge over the Danube at Turnu Severin in Romania—the longest bridge ever built by the Romans at 3,000 feet—it is estimated that slaves cut and hauled no fewer than one million rocks of about 2 feet cube, and dropped them in the Danube for the foundations. The time taken to complete a bridge was not a governing factor.

Foundations were built by simply throwing loose stones into the river until they were piled high above the water line or by using open cofferdams and timber piles. The Romans later evolved the art of pile-driving for forming bearing piles in the river bed and cofferdams, which were then enclosed by a foundation bed of concrete, to support the stonework of the piers.

All Roman arches were semicircular and were self-supporting, so that, if the enemy destroyed one arch, the others would remain standing. They generally built the piers to be one-third of the span, and that is why they were so massive.

The Romans knew how to build only the semicircular stone arch, but their mastery of this art has created the most noble structures ever seen or built in the history of civilization.

ABOVE: Pons Augustus as it looks today.

OPPOSITE: Pons Augustus, Rimini as it looked about fifty years ago.

Valentré, Cahors, France (1347) Pont Valentré is a fine example of a medieval fortified bridge. The bridge was built during the period when feudalism was fading out of the medieval world. This was the end of the era of the "Brothers of the Bridge," the Benedictine order of monks we met in Chapter 1. Merchants and public authorities were beginning to turn to local masons and craftsmen to build their town bridges, rather than wait many years for the Benedictine Order to build it for them.

Perhaps this might explain why it took 39 years to complete Valentré. The people of Cahors wanted a new bridge over the Lot, but, not wishing to break with the church entirely, they appointed the Bishop of Cahors as the nominal head of the project. The townspeople collected the money for the bridge by imposing a duty on all merchandising and commodities entering the city gates. It took from 1308 to 1347 to collect the money and to complete the bridge. Slowly and surely Pont Valentré was built, never more than two arches at time, so that building work could be easily adjourned without problems if the work was interrupted by war. (War was also the reason why medieval bridge builders made arches semicircular and stable and each pier an abutment, because if one arch was destroyed it would not cause the others to collapse.)

In regarding the bridge, the eye is drawn to the three towers rising 130 feet above the river Lot. The towers have tiled roofs covering their crenellated walls, from which fire could be directed at boats in the river as well as the approaching troops on the bridge. The six two-centered, arch spans have a width of 45 feet and have a recessed arch ring which accentuates the voussoir stones. The piers are 18 to 20 feet wide with triangular cutwaters, on both the upstream and downstream faces, that rise up to the roadway.

A series of holes can be found in the piers just below the springing line of the arch. Through these holes workmen pushed fir saplings until they jutted out at the either side of the pier. These supported a planked floor which provided a walkway for the workman, a resting place for the stone blocks, and a foundation for the centering. It is not usual to find piers so large in medieval bridges because the builder, being unable to calculate the actual stresses and load carried by the piers, used empirical rules to make sure they were safe.

Valentré also comes with a legend of the "Devil's Bridge," which was probably propaganda spread by the church about bridges that were not built by the Brothers of the Bridge. The legend states that the builder of Valentré sold his soul to the Devil after being discouraged by the slow progress made in the building. But the builder made one proviso: he would give his soul provided the Devil performed every task he was given by the builder. Work progressed very well until the builder had second thoughts about parting with his soul.

He thought he would catch the Devil out by asking the him to carry water in a sieve for the masons. No matter how fast the Devil flew, all the water had gone

BELOW: The medieval Pont Valentré over the river Lot in Cahors, France.

from the sieve by the time he reached the masons. The Devil admitted defeat and released the builder from his spell, but left him with this thought: "You have won," said the Devil, "but I will wager you that you do not boast of having had my gratuitous collaboration!"

When the bridge was completed and the builder was completing the central tower he found that a stone was missing in the northwest corner of the roof. He instructed the workmen to replace it and this they did, but, when they began work the following morning, it was missing. This went on day after day, with the builder replacing the missing stone only to find it missing the following day. The builder grew weary of the Devil's games and to this day—according to the legend—this stone is still missing.

Santa Trinità, Florence, Italy (circa 1570) Scholars and writers on arch bridges are unanimous in their vote for the best-engineered bridge structure of the Renaissance. For purity of form and its pencil-slim elliptical arches, the Santa Trinità over the Arno in Florence is without equal. The curve of the two arch spans defies analysis and has mystified engineers over the centuries when they have tried to explain its nature and how it was achieved. The arch consists of two curves, each resembling the upper part of a parabola, which meet at the crown at an obtuse angle, and the point at which they meet is concealed by a decorative cartouche or pendant. How did bridge builders in the sixteenth century know how to construct such a slender arch span, with the limited technology and crude mathematics that they knew?

RIGHT: The famous statues sculptured by Francaville, restored to the Santa Trinità Bridge after they were destroyed in the second world war.

ABOVE: The Santa Trinità Bridge over the Arno in Florence, Italy.

The curve is very shallow, and has a gradient of one in seven, which is startling when you consider that most arches built around this period had gradients of one in four at the most! It is also a breathtakingly elegant bridge.

In 1567 Grand Duke Cosimo I ordered a masonry arch bridge to replace the old wooden Santa Trinità over the Arno. Bartolommeo Ammannati was the Grand Duke's chief engineer and was asked to design and supervise the building. All sorts of theories abound as to how Ammannati arrived at the curve of the arch. There are no records to provide the answer, and, since there were no mathematical procedures laid down at the time, there seems to be only one answer. Ammannati designed it from aesthetic judgment, being a good draftsman and artist, although it is thought that Michelangelo had a hand in the profile.

Ammannati apparently drew two tangents as determinants. One was drawn vertical to the springing and the other slightly inclined to the horizontal at the crown of the arch. These tangents were then connected by a graceful curve, one for each half of the span. These meet at a slight angle at the crown, which is why Ammannati hid it with the cartouche. The Santa Trinità stood undisturbed for over 300 years, until 1944, when it was demolished by the Germans in World War II. It has been lovingly and painstakingly rebuilt and restored to its original glory.

The four famous statues that represent the four seasons, and which were added to the bridge in the seventeenth century, by the sculptor Francaville, have also been recovered.

Rialto, Venice, Italy (1591) "The best building raised in the time of the Grotesque Renaissance, very noble in its simplicity, in its proportions, and in its masonry." That is what John Ruskin wrote to describe the Rialto Bridge in Venice. For the first time in the history of bridge building, bridges were designed as architecture that was befitting the civic spirit of their community, to be graceful and aesthetically pleasing as well as being functional and well engineered.

The designer of the bridge was Antonio Da Ponte, a surname that means "bridge." It was a common name in Renaissance Italy. Not much is known about the hero of this story, who was a native of Venice and a curator of public works. The most remarkable fact about Da Ponte was that he was 75 when he won the contract to design and build the Rialto. His fame as a designer came to the notice of the Venetian Senate when he saved the Ducal Palace from complete destruction, after a serious fire broke out in the Palace in 1577. In spite of the dangers of collapsing timbers and molten lead spilling down from the burning roof, Da Ponte entered the flaming building and took charge of the situation. It was his courage and presence of mind that saved the palace. He also argued convincingly against Palladio's suggestions that the palace should be demolished and a new building erected after the fire.

Da Ponte saved for posterity one of the best examples of Italian Renaissance architecture at the same time winning for himself the distinction of designing all the repairs. When in 1587 the Senate invited designs for the Railto, Da Ponte won the contract, beating designs by Palladio, Michelangelo, Fra Gioconda, and other such eminent architects.

BELOW: Rialto Bridge, Venice, Italy.

LEFT: The central gallery and walkway of the Rialto Bridge.

BELOW: The central arch over the covered walkway.

The bridge stands today just as Da Ponte had designed it. In order to give free access to the canal boats, a single segmental arch covers the distance between the abutments. The arch has a clear span of 88 feet, and rises 20 feet above the springing. The total width of the bridge is 75 feet and it has a roadway down the middle, with shops on both sides and a footpath and parapet down each side.

Two sets of arches, six on each side of the large central arch, support the roof and enclose the 24 shops within it. It took three and a half years to build and kept all the stonemasons in the city fully occupied for two of them. According to the records 800 larchwood timbers were used to erect the falsework for the arch.

The foundation of the bridge created controversy, although it proved to be of very durable construction. Da Ponte used timber-bearing piles driven into the soft, silty canal bed by a mechanical hammer operated by four men, to form tightly packed rafts. The piled rafts were driven deeper into the canal bed the closer they approached the water. The tops were then trimmed off to form steps going down toward the water. The stepped rafts were timbered and surmounted with a brick platform built at the correct angle to receive the radiating courses of the abutment and spandrel masonry.

This was such a novel method of construction that many well respected engineers in the city raised doubts about the adequacy of the construction, causing the Senate to order the work to stop, pending an official inquiry. After expert witnesses were called in and Da Ponte had explained his method of construction, the Senate decided that the work was safe. The huge cut stones of the arch were now put in place supported on timber falsework built across the canal.

The bridge was opened in July 1591 and has stood the test of time remarkably well, even withstanding an earthquake without a crack, when Venice was shaken by a tremor shortly after the Rialto opened.

Pont de la Concorde, Paris, France (*c.* 1796) The greatest stone-arch bridge designs of the eighteenth century came from the inventive genius of Jean-Rodolphe Perronet, but, alas, his finest bridge, the Pont-Sainte-Maxence over the Oise, is no longer standing, having been destroyed in World War II. However, his most celebrated bridge, and the last bridge that Perronet designed, is still standing today. Perronet's discovery of the continuous action of thrust in segmental-arch construction, and his invention of the divided pier, completely

ABOVE: Building the arch supports of Pont de la Concorde over the Seine in Paris with timber centering. The water wheel, seen in the background, drained the coffer dam.

changed bridge design for all time. His arch bridges were more slender and spanned farther than any other. Perronet wished to make the Pont de la Concorde in Paris his last work, the most remarkable bridge structure ever built, with arches that were nearly flat and piers that were composed of just two Doric columns. But his vision was compromised by the ignorance of meddling officials and engineers who did not grasp the fundamentals of the design that Perronet had calculated and presented to them. He was 75 when he was invited to design the bridge and, because of his brilliance, there was no one with the intellect to appreciate his ideas. He was forced to increase the arch curvatures and to thicken the piers to satisfy the skeptics. The final result was a less slender bridge than Sainte-Maxience, but Perronet skillfully heightened the appearance of slenderness by introducing a balustrade instead of a solid wall for the roadway, which was the tradition for bridges in those days.

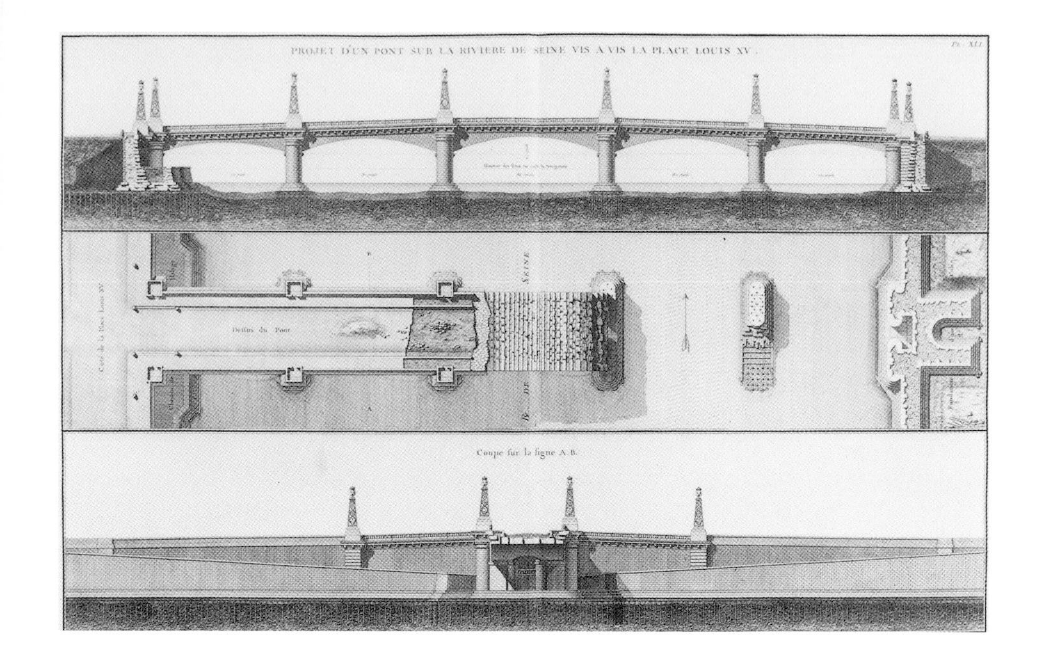

LEFT: Peronnet's drawings of the Pont de la Concorde.

BELOW: The Pont de la Concorde as it looks today.

Perronet was 80 when work started on the foundations of the bridge. They were of the usual design of wooden platforms on timber piles, about 6 feet 6 inches below water level. When the bridge was half built, the French Revolution broke out. The chaos that followed did not affect work on the bridge. Perronet oversaw everything and lived in a little house built at one end of the bridge. And when the Bastille was razed to the ground the stones provided a good source of masonry! When crossing the Pont de la Concorde, you are walking on history … the ghost of the legendary "man in the iron mask" probably walks there today … Many of the laborers who built the bridge belonged to the military force of the community of Paris, the formidable force that beat the National Guard and tore down the Tuileries.

The bridge has a noble and monumental presence. Its graceful lines and skillful construction make it an artistic triumph. It is clad in marble dressed with a classical ornamentation along its flanks. The decorative feature of the corbeled cornice is believed to have been inspired by the Pons Augustus in Rimini.

LEFT: The balustrade of the bridge is so reminiscent of the balustrade of the ancient Pons Augustus.

St Louis Bridge, St Louis, USA (1874) While Europe continued to build with wrought iron until the 1900s, American bridge designers took the bold decision to experiment with steel and to produce the first major bridges using it. The St Louis and Brooklyn bridges were the first bridges in the world to use steel as the principal structural material. Before this, steel was used only with wrought iron, as an expensive but efficient material for longer-span bridges. Steel was used for the main arch ribs on the St Louis Bridge, and for the cable stays and bridge deck trusses on the Brooklyn Bridge. Both bridges were under construction at the same time, but the St Louis was completed nine years before the Brooklyn.

James Buchanan Eades was one of the most gifted engineers of America in his day. You could say he was a latter-day Leonardo da Vinci, setting up many new ventures, and establishing many successful marine and industrial companies in St Louis during his lifetime. And, when he was approached to design a bridge over the mighty Mississippi, he designed the most daring structure ever seen.

Eades had never built a bridge before, so why did the city fathers and bridge sponsors trust him with the task? For 20 miles above St Louis, where the Missouri empties into the upper Mississippi, it turns into a deep and seething torrent, which is subject to great changes, both seasonal and tidal. The speed of the river can vary from 4 feet per second to 12 feet 6 inches per second in flood, and has a tidal range of 41 feet from low to high tide. It's a brute of a river because, as the river flow increases, sand and mud deposited on the river bed is churned up and carried downstream; as it slows, sediment is deposited on the river bed.

The movement of the river bed was understood by Eades better than any other engineer in St Louis, through his extensive studies and personal observations as a riverboat captain, as a designer of iron-clad gunships during the American Civil War, and as a salvage operator on the Mississippi.

Even when the Missouri legislature agreed to grant a charter in 1864 to a consortium calling themselves the "St Louis and Illinois Bridge Company" led by Norman Cutter, there were delaying tactics by rival factions and objectors to the bridge, principally from the ferry boat companies. Meanwhile, more winters were to pass, with the Mississippi freezing over and ice floes scouring the river bed and presenting a hazard to shipping. At times the river was too treacherous for navigation, cutting off communication and traffic for weeks between the east and west of the city.

BELOW: The bridge deck being built up from the arch.

Finally, on August 20, 1867, actual construction was started, but even when the corner stone of the west abutment was laid in February in 1868 a rival bridge group confusingly calling themselves the "Illinois and St Louis Bridge Company" tried to hijack the bridge project from Eades and his backers. The dispute was settled in March, 1868, when both bridge companies merged and Eades was made chief engineer.

In trying to build a cofferdam for the western abutment, the construction team had to drive through 60 years of scrap dumped in the river. Eades had known that there were at least three burned-out steamer hulls, the wreckage of four barges, anchors, chains, and all sorts lying above the bedrock he wanted to reach. All this made it extremely difficult to make a watertight cofferdam. The pieces that could be hauled out of the water easily were removed, but much of the stuff was buried deep under the silt and mud. He designed a gigantic wooden chisel with a steel blade to cut and smash a way through this underwater scrap yard. Eades wanted to found the cofferdam on bedrock, knowing it would be folly to sit the foundation on the gravel layer above it, which could be scoured by the river in flood.

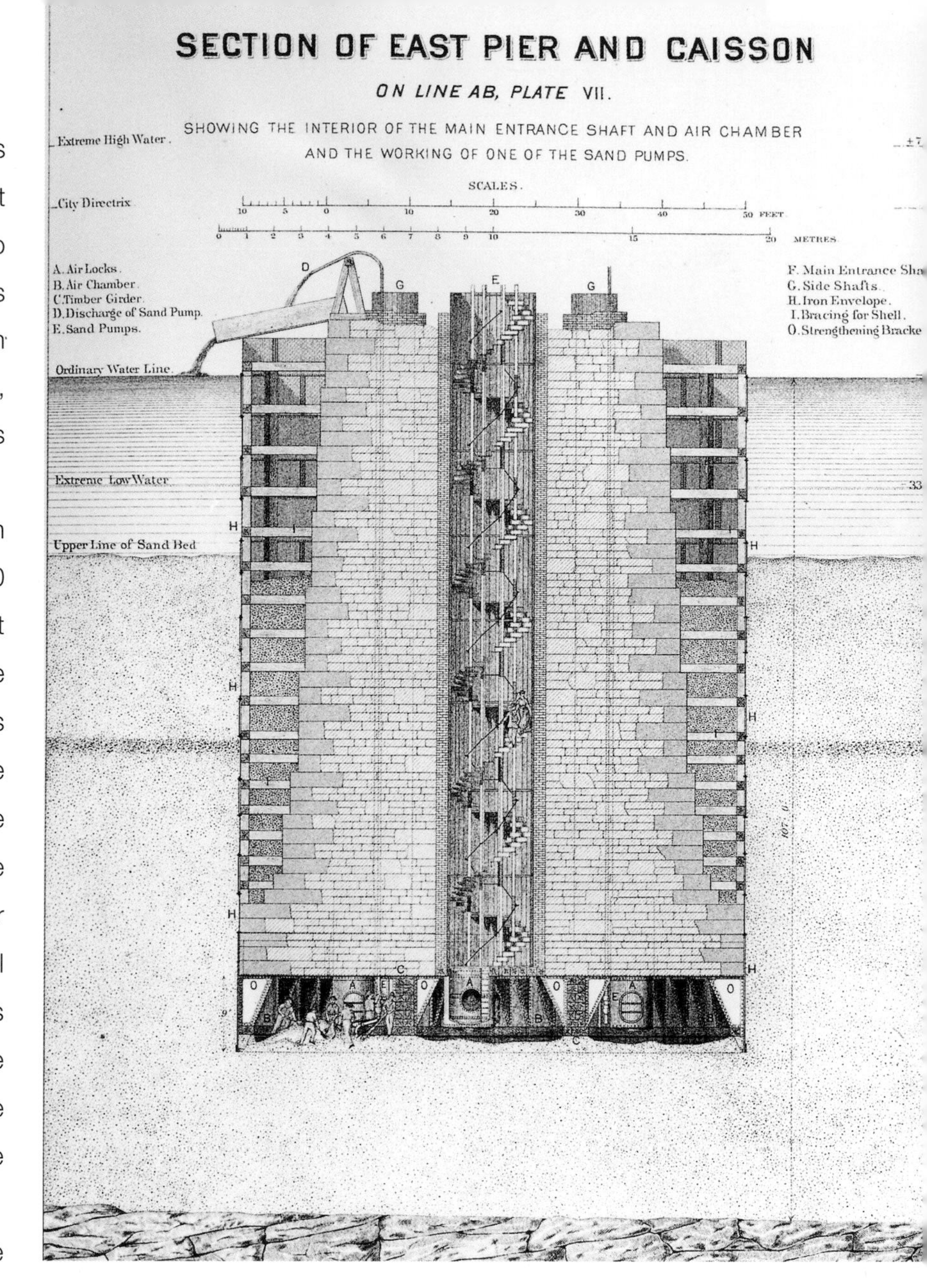

ABOVE: Diagram showing the caisson of the east pier.

Having driven the west abutment and built up the masonry structure within the cofferdam, Eades now faced the harder challenge. The east abutment was some 130 feet down to bedrock—60 feet of water and 70 feet of sand and silt—and was deeper than any caisson had been sunk. He studied the reports of the ingenious pneumatic air caisson developed by Brunel at Saltash for the piers of the Royal Albert Bridge and crossed to France to discuss pneumatic-caisson construction with Monsieur Audernt, who was sinking one for a bridge over the Allier. On returning to St Louis he postponed the east abutment until he had built the two-pier foundations in the river, using this technique, as they were shallower.

The caisson was a huge rectangular box made of wood and sheeted in iron panels and stiffened with girder plate. Inside the caisson there were stairwells leading down to the working chamber, a sealed compartment, which was 9 feet high and bottomless. Other airtight shafts were built for removing excavated material out of the working chambers. As the masonry pier within the cofferdam was built up from the roof of the working chamber, the caisson would start to sink into the water until it reached the river bed. To stop water filling the working chamber, air was pumped in to equalize the water pressure.

Once the caisson reached the bottom of the river bed, down the ladders and through the airlock chambers went the workmen. The air pressure inside the chamber was increased as the caisson was excavated slowly down to the bedrock. This could take many, many months depending on the difficulty of alluvium to be removed. Eades had devised a sand pump, the first of its kind in the world, to dispose of the river excavation, rather than use the traditional system of buckets on a rope, which was much slower. Unfortunately, not much was known about decompression sickness or, as it is sometimes called, caisson disease or the bends, caused by working under high pressure and decompressing too quickly. On the east and west pier caisson, there were 91 cases of the bends, with 13 deaths and two persons crippled for life. Eades sought the advice of the best medical brains, and a Dr Jaminet was hired as medical adviser. Jaminet suggested a slow decompression and cutting down the working shift to two hours at 32 pounds pressure; and to one-hour shifts when under 34.5 pounds pressure. There were few problems when this procedure was adopted, although the decompression rate used was still too fast at 6 pounds per minute. The safe figure is 1 pound per minute, but this was not discovered until many years and many bridges later! For the east abutment, which was much deeper, there were surprisingly few injuries, after the working restrictions imposed by the doctor were followed.

Eades next had to present to his promoters and critics the economic and structural justification for building a three-span, steel-arch bridge on a scale never seen before. In his defense of the feasibility of the steel arch, this is what he wrote:

> "In 1801 the great Scottish engineer, Thomas Telford proposed to replace the Old London Bridge with one of cast iron, having a span arch of 600 feet … For forty years this remarkable man continued to enrich Scotland and England with some of the most stupendous and successful triumphs of engineering skill to be found in Great Britain … Surely the recorded judgment of such a man as Telford when sustained by the most eminent men of his day, asserting the practicality of a cast iron arch of 600 feet span in 1801, furnishes some "engineering precedent" to justify a span of 100 feet less in 1867."

BELOW: The St Louis Bridge spans the mighty Mississippi (1874).

The fabrication of the steelwork was awarded to the Keystone Bridge Company, one of the finest bridge companies in the USA. Eades had devised test apparatus to assess the structural quality of the steel plate that was rolled to form the tubes of the main arch. After teething difficulties in achieving the grade and strength of steel required, scaled-down sections of the arch were made and tested. These showed that the steel tube for the arches would be more than adequate to resist the stresses imposed on them on the actual bridge.

The three great arches were erected without falsework, built by cantilevering the arches out from the piers toward the center of the span. Temporary towers were erected on the piers to support the tie-backs. The tie-backs or cantilever cables were made of steel bar an inch thick and 6 inches wide. As the arches cantilevered farther and farther out over the water the next problem arose. How to ensure that the two halves meet in mid-span? To ensure this happened Eades built each section bigger by a small factor, which would produce an overlap at mid-span. And if this did not work he designed closure tubes to slide over both sections at the crown.

On September 14, 1873, on the first attempt to close the inner ribs of the western span, there was a gap so small that the closing tubes could not be fitted. A loan of half a million dollars depended on closing the first span by September 19, five days' time. This money was vital to keep the project running. By cooling the steel it was thought it might be possible to close the gap, so the arch rings were packed with ice, which was placed in wooden troughs built around them. Fifteen tons of ice was placed in the troughs on September 15, and by sunrise the following day the gap was still five-eighths of an inch. Forty more tons of ice was placed the next day, but by sundown the arch had closed no further.

With no prospect of cooling weather in the next few days, the ice poultice was abandoned and a special adjustable closure tube that had been designed by Eades was hurriedly fabricated. On September 18, within a day of the deadline, the arch was closed. By December the inner ribs of the central and eastern arches' spans were also closed. It was fairly straightforward now to finish the outer ribs and to build the roadway and railroad deck structure.

On July 1, 1874, Eades organized a public show of the strength and integrity of the bridge, halting 14 heavy steam locomotives—seven on each track—over each span and then sending all 14 in single file over all the three spans.

And on July 4, 1874, Independence Day, the city of St Louis celebrated the opening of the bridge in magnificent style. A mammoth procession paraded through the streets of the city, crossing and recrossing the bridge. In the evening a fantastic fireworks display launched from the roadway of the bridge lit up the night sky and the triumphal arch constructed near the bridge portal that was topped with a portrait of James Buchanan Eades bearing the inscription "The Mississippi, discovered by Marquette, 1673; spanned by Captain Eades, 1874."

ABOVE: The truss section of the railway platform below the roadway deck of the St Louis Bridge.

Garabit Viaduct, St Flour, France (1884) In France the spread of the railroads and the growth of industry were slower than in the United States and Britain. To some extent this was attributable to the remoteness of the raw materials from the towns and cities. Many important mineral deposits were located in the high, barren plains of the Massif Central, which meant opening a railroad before minerals could be commercially exploited and sent to the large towns of Lyon and Limoges. Alexander Gustav Eiffel had been working on the rail network for many years and had built many fine iron viaducts across deep gorges. He had developed methods of designing pylons, towers, and truss girders, to withstand the high winds that funneled down the valleys. Eiffel's designs were capable of resisting the wind forces, and the vibrations that were caused by it. He made sophisticated measurements of the wind force and wind direction and then studied the effects this would have on the structure. The stiff lattice trusses that he designed, with their large open spaces and minimum wind resistance, were Eiffel's solution to building high-masted structures and long-span bridges across steep valleys.

ABOVE: The Garabit Viaduct in St Flour. The height of the arch above the water line is higher than the Eiffel Tower.

The Garabit Viaduct over the Truyère at St Flour is an important structure because it marks the changeover from iron to steel in bridge construction. The two-hinged-arch concept that Eiffel designed for Garabit was to become the standard design of steel arches that were to follow.

But Garabit is not a steel structure: it is built from wrought iron. There were difficulties in the manufacturing quality of steel at the time, and it was more expensive, so it was not the best material to build with. Eiffel's 530-foot, wrought-iron, parabolic arch supports the 1,850-foot-long truss of the railroad, some 400 feet above the river Truyère. The arch is narrow and deep at the crown to carry the railroad trusses, while at the supports it is wide and shallow. The ends of the arch rest on hinges, which allow for the expansion and contraction of the bridge.

Eiffel pioneered the cantilever construction method for his metal-arch bridges to eliminate the high cost of building falsework below the bridge. The half-sections of the Garabit arch were built out from each abutment and held in position temporarily, using wire stays. Garabit was the longest and highest arch bridge in the world when it was built, exceeding Eiffel's arch truss over the Douro in Oporto by 50 feet. It was finished in 1884 and was the last bridge that he built.

ABOVE: A support pier and the truss of the railway deck.

RIGHT: The changing profile of the arch is clearly visible—from wide and thin at the base to deep and narrow at the crown.

FOLLOWING SPREAD: Close up of the open structured bracing of Gustav Eiffel's wind resistant arch.

George Washington Bridge, New York, USA (1931) Like the Empire State Building, the George Washington has come to be regarded as one of America's greatest constructions. Crossing the Hudson at its widest point, with a clear span of 3,500 feet and steel towers rising high above the water level, the George Washington Bridge smashed the longest span record by 1,700 feet—a margin that is as much as the actual span of the Firth of Forth! With massive spans like the George Washington, the suspension bridge displaced the cantilever-truss bridges as the accepted type of long-span bridge of the future.

In the George Washington Bridge we see the culmination of the many advances in suspension-bridge construction since the Brooklyn Bridge was finished in 1888. It did have flexible steel towers, which were first pioneered on the wonderful Manhattan Bridge and made it possible to reduce the weight of steelwork. It incorporated advanced cable-spinning technology to lay the pair of 3-foot-diameter cables that make up the suspension system on each side of the bridge deck. The cables were spun in 209 working days with a labor force of over 300 men. In just seven minutes six loops were passed from end anchorage to end anchorage—a distance of a mile—and in an hour 100 miles of wire were spun. It required 217 loops to form a strand and there were 61 strands to make up each 3-foot-diameter cable. The accurate spinning of 107,000 miles of wire, which could encircle the equator four times, was an unprecedented feat of construction.

The tower steelwork was floated to the piers in 50-foot-long sections, and it needed 12 sections to complete each 635-foot tower. The sections of steelwork, after riveting and fabrication, were erected by huge derricks very rapidly. Workmen handled a million white-hot rivets, which had to be flung through the air hundreds of feet above the river, caught in buckets, and then driven into the steel with a pneumatic hammer, to complete all the joints of the tower sections. When the 180-ton saddles for the cables were in position on the top of each tower, it is estimated that 20,000 tons of steel had been used to build both towers.

The large caisson sunk for the west tower foundation was the largest cofferdam ever constructed. The east tower foundation was built on the shore and the massive cable anchorages were formed in the basalt rock on each bank, pouring concrete into a vast hole 220 feet by 209 feet by 130 feet deep blasted out of the rock. On such a natural site for a bridge, there was little that could go wrong.

The bridge was built in four years and opened on October 25, 1931, and is believed to be the only bridge ever built that was finished a year ahead of schedule! The bridge engineer was the great O.H. Amman and the architect was Cass Gilbert, the designer of the gothic Woolworth Building. The steelwork to the towers had been designed to be covered by concrete and granite panels, but, as the skeletal profile of the tower soared skyward, the natural beauty of the fabricated section fascinated and excited onlookers so much, that the bridge authority decided not to cover it, saving quite a tidy sum of money in the process.

LEFT ABOVE: A dramatic view of the distinctive steel lattice work of the George Washington Bridge tower.

LEFT BELOW: The bridge seen through the tower on the Manhattan side.

ABOVE: George Washington Bridge, New York, USA (1931).

Firth of Forth, Scotland (1890) This was the most massive, monumental cantilever-truss structure ever built. Was it a masterpiece of engineering or a metal dinosaur that should never have been built? Some say it was so dated in its engineering that it cannot be hailed as a masterpiece of bridge design. But, on seeing it, one cannot fail to be moved by the sheer scale and muscular proportions of the trusses, which create such an awesome presence across a cold, weather-beaten landscape, that it evokes adulation and inspires admiration by everyone.

When it was finished, the Firth of Forth was the longest-spanning bridge in the world, beating the Brooklyn Bridge span by 100 feet. It established the cantilever-truss bridge as a serious competitor to the suspension bridge. There are still many fine cantilever-truss bridges standing today, but none were to match the spans of the Forth. The tragic collapse of the first Quebec Bridge expunged the growing trust in the efficacy of long-span cantilever construction. The honeymoon period for the long-span cantilever bridge was therefore short-lived.

The Firth of Forth Bridge would never have been built as a cantilever-truss construction if the Tay Bridge had not collapsed. Thomas Bouch's proposal for a long-span wire-cable suspension bridge had secured authorization from Parliament and had received funding, and work had actually commenced when, on December 20, 1879, the Tay Bridge, which he had designed, collapsed. Bouch was found negligent and of poor engineering judgment in the design of the main portals, and a year later he was asked to resign as designer of the Forth Bridge.

BELOW: Completing one of the balanced cantilever span sections.

ABOVE: Building one of the main piers of the cantilever span.

New proposals for the bridge were invited from the consulting engineers, Sir John Fowler, W.H. Barrow, and T.E. Harrison. From them emerged a design, based on the continuous girder truss but which had definite breaks or hinges at points of low stress. These hinges transform the bridge into a series of cantilever spans and suspended spans which made the bridge easier to assemble, although it required more steel. The original proposal was later modified and improved on by Fowler's junior partner, Benjamin Baker. The Pratt truss sections for the cantilevers were not as stable under construction as the double triangular form that Baker later introduced. He also chose to have a smaller number of large sections to form the cantilever, rather than a large number of smaller structural members.

For the principal compression members of the towers, Baker designed four 343-foot-long tubular struts, 12 feet in diameter. A circular shape is more efficient for a compression member and is less prone to buckling than a flat-plated surface. It would cost more, but Baker argued that this would be outweighed by the structural advantages.

The erection and building of the cantilever spans, the tower columns, and caisson foundations was relatively uneventful. The site chosen was ideal, because the river bed where the piers were founded consisted of hard, durable rock. The foundations for each of the three towers were circular granite piers, which were battered for increased stability against the wind.

The whole structure consists of masonry-arch approach viaducts—one on either side of the main cantilever structure. The main bridge comprises two anchor spans, flanking three huge towers of steel, from which the cantilever-truss arm projects to carry the two suspended spans. There were few tragic accidents for the enormous labor force of 4,500 men who were employed to build the bridge—thanks to careful planning and safety measures. The steel, after it had passed through the plating shops, was coated with boiling oil. Granite, rubble, and sand were ferried by water and brought directly to the site. Four steam launches and eight large steam barges transported the materials, which amounted to some 140,000 yards of masonry and 58,000 tons of open-hearth steel.

On March 4, 1890, the bridge was opened. For generations the Forth Rail Bridge remained the longest-span bridge in the world. It represents the first large-scale use of open-hearth steel in bridge construction and was for over 50 years the only long-span bridge over which express trains can travel at 60 miles an hour.

FOLLOWING FOUR PAGE SPREAD: Panorama of the Forth Rail Bridge across the Firth of Forth, Scotland.

BELOW: Detail of one of the tubular steel truss members and bolted connections.

Plougastel, Elorn, France (1929) This bridge represents the genius of one man's inventiveness, the exploitation of the plastic properties of concrete, the daring use of prestressing in concrete and an ingenious solution to building a multiple-arch span across a wide river.

Eugene Freyssinet's bridge over the Elorn at Plougastel was the finest bridge he had ever built and one of the truly great achievements in the art of concrete-bridge engineering the world has ever seen. This was the dawn of the age of concrete so masterfully heralded and understood by the bridge designs of Robert Maillart, but it was not until Eugene Freyssinet arrived that concrete-bridge engineering was taken to the limits of its capability. Freyssinet, the father and

BELOW: Plougastel Bridge as seen before a second bridge was built nearby.

inventor of prestressed concrete, was not satisfied with just making this discovery in his lifetime: he went on to exploit its unique properties, designing elegant shell roof structures, long-span exhibition halls, aircraft hangars, and bridges.

It was 1925, and cement technology was in its infancy: ready-mixed concrete, super-strength concrete, and pumped concrete had not yet been developed. Everything was mixed on site by crude drum mixers fed by workmen shoveling the materials into them—the large aggregates, the sand, and the cement. There was simple apparatus for weighing out the materials—the cement was supplied in bags, which had to be split open. On a windy day that fine grey power would get blown everywhere.

And yet, when Freyssinet submitted his winning design for the competition to build a bridge at Plougastel, he had the audacity to believe that he could produce a high-strength, rapid-hardening concrete with such crude technology. Using about 755 lbs/cu yard (450 kg/m³) of cement, he was able to achieve it! You need careful control of the mix proportions, computer-controlled weigh-batching, and specially formulated admixtures to make high-strength concrete from a ready-mix plant today. Freyssinet's proposal was akin to giving a surgeon a bread knife to perform a delicate operation on an eyelid—but he did it!

He needed high-strength concrete to construct the three identical concrete arches, each spanning 600 feet over the Elorn. The bridge was the longest span in France, beating the Garabit Viaduct, and for a short while it was also the world's longest concrete arch. The three identical arch spans have a cross section comprising a three-celled box, which is 31 feet wide and 16 feet deep at the crown. Spandrel walls built up from the arch support a double-deck truss which carries the roadway and a single rail track. The upper deck of the truss forms a 20-foot-wide roadway bordered by a 3-foot sidewalk on each side. The arch was concreted and reinforced in three stages. First, the slab for the lower chord was cast, then the box section walls, followed by the slab for the upper chord. The high early strength of the concrete was used to enable the prestressing to be applied after three days, before the concrete could shrink and crack. The other brilliant achievement in engineering the Plougastel Bridge is the story of how it was constructed. This is where Freyssinet saved money and time, and was able to beat the price of his competitors' designs. The bridge itself needed about 32,300 cubic yards of concrete along with forming materials, centering, and reinforcement. The difference in high and low tides can be as much as 26 feet in the Elorn, and, with wind speeds as high as 100 feet per second, it makes construction difficult in the river.

Instead of building the falsework for the centering across the river for all three spans in the traditional way, Freyssinet made a pontoon and floated out the centering for one arch only. This was removed and reused for constructing the other two arches in turn. The centering for the arch required only 260 cubic yards

BELOW: 1. Centering being fabricated on the shore **2.** Manoeuvring the centering into position for the second arch span **3.** Towing the centering into the Elorn estuary **4.** The second arch span being built, with the completed first arch span awaiting the start of the vertical piers and the bridge deck

1

3

of timber and was built on the Plougastel side of the river. A timber segmental arch was constructed with each end seated in a hollow concrete pontoon. The ends of the centering were tied together by cables whose tension could be regulated at both ends by hydraulic presses. Exploiting the tide, the concrete pontoons and the centering were towed out into position for the first arch on April 2, 1928. Once they were in position, the hydraulic presses lifted the centering to spring it to the correct level, and to the correct alignment with the help of the hangers at both ends of the centering. By August 7, when the concrete of the first arch had fully hardened, the centering was eased free and floated into the second opening. On January 19, 1929, it was floated into the third opening.

2

4

Sydney Harbour Bridge, Sydney, Australia (1932) The Sydney Harbour Bridge is a monument to Gustav Lindenthal's great Hell Gate Bridge. Lindenthal was the man who proved beyond question that a steel arch could be a thing of real beauty. Sydney is situated in a superb natural harbor—often claimed to be the most beautiful natural harbor in the world—but, because of it, only the south side of the city had grown, as the north side could be reached only by ferry or a circuitous ten-mile coastal road. Many proposals for a bridge had been made over the years, but nothing was workable since the ferries were providing an efficient service and the cost of a bridge was unjustifiable.

ABOVE: Steel truss of the Sydney Harbour Bridge. 50,300 tons of steel were used in the making of the bridge.

Then in 1922 John Bradfield, the chief engineer of the public-works department, published his proposal for a cantilever bridge across the harbor. But, before he finalized his design, he decided to go on a fact-finding mission to the United States and Europe, to study the world's long-span bridges. He was particularly impressed by the Hell Gate Bridge and upon his return to Sydney revised his design and copied the steel arch.

Sometime in 1923, after many bridge companies and engineers of the world had been invited to submit their design and tender proposals, the firm of Dorman Long were awarded the contract to build the Sydney Harbour Bridge. Not surprisingly, it was based on the steel arch that Bradfield had proposed, rather than a suspension bridge. There was a view that, with rock so near the surface, a suspension bridge would have been the better choice, but the competitive tender of Dorman Long suggested otherwise. Aesthetic grounds were cited for the selection of the steel arch, but it is also likely that a deciding factor was the ambitions of the Australians to boast the longest-arch span in the world.

The Sydney Harbour Bridge, like the Hell Gate arch, carries four railroad tracks, but in addition it carries a wide roadway and sidewalks. The dead weight of the span is 57,500 pounds per linear foot, while the Hell Gate was 52,000 pounds per linear foot. However, Sydney Harbour, built in 1932, spans 1,650 feet between abutments, while the Hell Gate, built in 1916, spans only 977 feet in comparison.

To complete the bridge 50,300 tons of steel was used, 37,000 tons of which was contained in the main arch. Special steel-fabrication yards were built close to the site of the bridge on the north bank. Panels of trusses were accurately assembled in 60-foot lengths. After riveting and painting, they were loaded onto barges and transported to a position directly under the bridge and lifted into place from the barge, by a 120-ton overhead traveling crane.

The erection of the main arch span was one of the greatest engineering feats of its time. The two halves of the arch were built out from each abutment, with each half held in position during erection by 128 steel cables, each 2.75 inches thick. These cables were attached to the end post of a truss, passed through a U-shaped tunnel cored into the solid rock foundation on the bank and then attached to the end post of another truss. Two traveling cranes were mounted on the top chord of the truss to carry the material for erecting the steelwork. Each crane was electrically operated and had a lifting capacity of 120 tons and weighed 565 tons. Between them, they erected all the steel for the steel arch.

To close the span the half-arches were lowered at the center, by gently slackening the anchorage cable at the links at the top of the end posts. Each cable was let out at both ends simultaneously. A team of six men on each half, in constant telephone contact, carried out the task, working two 12-hour shifts. It took five days to reduce the center gap to just 8 inches. The lower chord of each half was then connected by an 8-inch pin enclosed by a forged saddle fitted to the chord. The remaining tension in the cable was released very gradually, while the steelwork over the crown of the top chord was then assembled, joining both halves.

Now all that remained was to place the hangers from the main arch, the cross girders, and the bridge deck below. This work commenced from the center of the span with the cranes traveling backward toward the abutments. In May, 1933, the last piece of steel was placed for the bridge deck and the cranes were dismantled. The granite-clad towers that mark the ends of the arch and the intersection with the approach viaducts are an ornamental feature to convey the impression of mass. They do stabilize the arch but only at the springing line at the base of the tower.

ABOVE: Sydney Harbour Bridge, Australia (1932).

Early in 1932 Sydney Harbour Bridge was opened and all the citizens of the city celebrated the completion of this monumental achievement, but they could not rejoice in the claim that this was the longest arch in the world. Four months earlier the Bayonne Bridge in New York was opened. It was 2 feet longer!

RIGHT AND BELOW: Two views of Sydney Harbour Bridge, through the lens of photographer Grant Smith.

The Golden Gate Bridge, San Francisco, USA (1937)

The Golden Gate Bridge is one of the acknowledged wonders of the modern world, and the universal symbol of the modern suspension bridge. It stands for achievement, progress, and breathtaking imagery—it is the logo of a city that is known throughout the world, even though many people have never been to San Francisco or know precisely where the bridge is situated. What is so great and so magical about this bridge? Is it the tall red towers, the mile-long cables that span the bay, or the setting of the bay with the city in the background?

ABOVE: The "Span of Gold" across San Francisco Bay.

The span of the Golden Gate is almost unbelievable at 4,200 feet between the towers. The distinctive stepped-back towers soar 746 feet in the air and are the tallest cable masts in the world. On pure engineering excellence the Golden Gate cannot be regarded as a very innovative or pioneering structure in its day—it just happens to be big. The sag of cables is excessive, the truss deck is very plain, and the architectural modeling of the towers by Irving Morrow is very affected, although it works extremely well.

Many engineers and bridge historians consider the 2,100-foot span of West Bay Bridge situated farther along the estuary to be a finer structure. But the Golden Gate has captured the hearts of the public and has become an institution, a monument to the enterprise and dynamism of San Francisco and the USA. It would be unthinkable to anyone today for such a bridge not to have been built in such a magnificent setting. And yet it did take a long time for the money to be found and the economics to be justified, before this bridge was built.

The need for a crossing between San Francisco and Saulite was not so pressing as the Transbay link between San Francisco and Oakland, which ferried millions of people every year across eight miles of open sea. The Transbay bridge, which included the West Bay and East Bay bridges, was built at a cost of $79 million between 1933 and 1936 and is still the longest high-level bridge in the USA at 43,500 feet.

When a scheme submitted by the consulting engineer J. Strauss to the city engineer and estimated to cost $27 million, it was received with some surprise. The bridge that Strauss had proposed was of a cantilever-truss design, supported

by suspension cables—it was also considerably cheaper than the budget of £100 million for a suspension bridge. Strauss argued that, for such a long span, a cantilever-truss construction like the Forth Bridge was too heavy, and a suspension bridge was not going to be rigid enough in such a windblown corridor, which also needed deep foundations in the sea. But some doubted that the design was worthy of a bridge that was to become the longest span in the world. Nevertheless, the city authorities were enthusiastic and canvassed support from the nearby counties and major industries to secure the necessary legislation to set up the Golden Gate Bridge Authority. An "opposition committee" led by prominent businessmen, taxpayers, and leading engineers was also formed to make counterclaims that the bridge was not needed and was impractical to build. After many years of political wrangling and arguing, it was left to the people of San Francisco to decide the fate of the bridge. On November 4, 1930, the public voted for the bridge to go ahead and construction finally began on January 5, 1933. In the intervening years Strauss's hybrid design was revised to a suspension bridge after consulting with Amman and other experts. It is the design that we see today.

ABOVE: The deep truss deck of the Tagus Bridge, seen here, was copied from the Golden Gate Bridge.

The most difficult part of constructing the Golden Gate Bridge was the pier foundations. This seems to ring true for all bridge types, from the smallest spans to the very largest bridges in the world. The pier on the west side encountered no difficulty, but the one on the other side, being in open sea, was unprotected from the elements and was vulnerable to the hazards of oceangoing ships. An access trestle bridge was built out to the pier 1,000 feet from the San Francisco shore. This is how men and materials for constructing the foundation and the superstructure of the pier accessed the site.

Not long after the trestle was built a ship was thrown off course in a fog and crashed into it, doing a lot of damage. And not long after that, in gale-force winds, 800 feet of it was carried away by the sea. It was rebuilt, and this time it was anchored to the sea bed and stayed in position for the duration of the construction, without further incident. When the caisson for the east pier had been towed into position inside a large concrete fender, it started to bob around like a cork on water under a heavy sea swell that developed through the night. The fender was there to stop ships colliding with the pier when the bridge was built, and to protect the caisson under construction, but ironically the caisson was now threatening to batter the fender wall to pieces. The caisson was successfully towed out of the fender and the fender ring was enclosed and used as a cofferdam.

ABOVE: Sailing through the hangers of the suspension cable!

OPPOSITE: The distinctive stepped back towers of the Golden Gate Bridge rise 746 ft.

There were few serious incidents during the construction and it seemed that the Golden Gate Bridge was going to be blessed with good fortune. There is a saying among bridge workers that "the bridge demands a life"; loosely interpreted, that works out at one death per million dollars spent on the structure! From the start of the work only one life had been lost—that was until February 17, 1937, near the end of the contract. One of the scaffolds erected by the bridge paving contractor gave way, carrying with it 12 men and 2,000 feet of safety net, which was put there to stop men falling into the sea.

On May 27, 1937, a week of celebrations inaugurated the opening of the bridge. During this week the opening ceremony took place, at which the designer J. Strauss presented the bridge to the city. The nightly event of illuminating the bridge by floodlight was a great spectacle and the vermilion paintwork gave the structure the nickname of the "Span of Gold."

Severn Bridge, England (1966)

This is regarded as the world's first modern suspension bridge where the heavy truss deck—universally adopted after the Tacoma Narrows collapse—was abandoned in favor of a sleek aerodynamic box-girder design.

The bridge design was started in 1961 by a joint team of bridge engineers, Freeman Fox and Partners and Mott Hay and Anderson, with architectural consultancy provided by the Sir Percy Thomas partnership. The teams were appointed to design two of the longest suspension bridges in Britain and Europe at the time: the Forth Road Bridge, located just upstream of the famous rail bridge in Scotland, and the Severn Crossing near Bristol in the southwest of England.

Suspension-bridge design the world over had taken fright at the collapse of "Galloping Gertie," the nickname given to the first Tacoma Narrows bridge in Washington State, which began to twist, sway, and finally collapse under a moderate wind of 40 m.p.h. For the next quarter of a century all major suspension bridges were massively constructed with stiffening trusses, with older bridges like the Manhattan and the Golden Gate being retrofitted with deeper stiffening trusses.

The Forth Road (suspension) Bridge was designed conservatively with stiffening trusses, but the Severn, with a span of 3,240 feet, was formed with an innovative and radical streamlined deck section that minimized wind resistance and allowed huge economies of scale and material cost.

BELOW: Construction of the aerodynamic bridge deck of the Severn Bridge.

As early as 1953 the German bridge engineer Fritz Leonhardt had applied for a patent for a flexible stiffening truss which was suspended from only one supporting cable, and an A-frame cable tower, with rows of hangers arranged in a zigzag fashion, so that every four met at a point. This hanger arrangement helped to stabilize the suspension cable and bridge deck from wind oscillations. His later unsuccessful but radical competition design for the 1960 Tagus Bridge in Lisbon created quite an interest in the bridge world. Encouraged by the saving in cost that such a scheme potentially offered, Gilbert Roberts, the senior partner of Freeman Fox, sought Leonhardt's advice on streamlining the bridge deck of the Severn crossing and the feasibility of incorporating zigzag hangers.

LEFT: The stiff truss section of the Forth Road Bridge, which was built just before the Severn Bridge.

BELOW: A section of the radical aerodynamic bridge deck.

Wind-tunnel tests carried out by Freeman Fox proved conclusively that the combination of a streamlined deck with two supporting cables and diagonal hangers was just as effective for wind damping as a stiffening truss deck. It was the best way of dealing with the resonating, wind-induced oscillations, which had caused problems on the Forth Road Bridge, when the 512-foot-high towers started to sway.

The Severn's swift tidal flow created huge problems in building the foundations for the bridge pier. On the west pier workmen were able to work on the base only during two 20-minute periods a day, at low tide. After the towers were built and the cables spun, the 60-foot-long prefabricated "aerodynamic" metal deck sections were floated down river from the workshops and lifted by the hangers into position.

As well as a radical redesign of the deck, the towers of the Severn Bridge were designed as hollow box sections with internal stiffeners in the leg section, braced by deep portals, one immediately beneath the deck and two above it. This was a much lighter construction than designs based on the American cellular-plate construction, and, although the towers (445 feet) were 70 feet shorter than the those on the Forth Road Bridge (512 feet), they used only half the amount of steel.

The Severn Bridge was opened to traffic in September 1966, five years after work had started. The Severn's revolutionary deck structure and tower fabrication have been adopted on many subsequent suspension bridges, notably the Humber Bridge in the UK, which became the world's longest suspension bridge in 1981 with a span of 4,624 feet.

BELOW: Recent pictures of the Severn Bridge, taken after the bridge had been repaired and repainted.

Sunshine Skyway, Florida, USA (1986) Disaster strikes on May 9, 1980

A thunderstorm had formed to the west of the Gulf of Mexico and was moving toward Tampa Bay. It was early morning on May 9, 1980, as Captain John Lerro, a member of the elite Tampa Bay pilots' association, was sent out on a 55-foot launch to bring in an empty cargo freighter, the *Summit Venture*, before it entered the tricky channels in the bay and passed under the Sunshine Skyway. It was just another routine job for Captain Lerro. He had piloted many big and small freighter ships successfully through the shallows and the sharp dogleg channels of the bay.

BELOW: The freighter "Summit Venture," having crashed into one of the main piers of the original Sunshine Skyway bridge in May, 1980.

BOTTOM: "The end of the road" literally—the collapsed cantilever truss span.

A mist, then a drizzle, followed by hard driving rain, swept very suddenly across the ship and the calm bay without warning. On the deck of the *Summit Venture* three lookouts kept watch for the all-important marker buoy that would tell Captain Lerro where to turn in the channel. Visibility was lost in the blackening sky and the colorless sea as rain lashed the deck. The ship was still moving steadily ahead, although by now the crew and pilot could see nothing but a screen of rain.

Suddenly Captain Lerro saw the Sunshine Skyway bridge looming ahead. He shouted the orders: "Hard to port—let go the anchor—ram the engines full Eastern!" But it was too late. The *Summit Venture* smashed into the tall support of the main span. As the ship hit the bridge, concrete and steel came falling down, some of it landing on the bows. Not only did the bridge deck collapse: six cars, a bus, and a truck fatefully traveling over the main span plummeted 150 feet into the sea. Captain Lerro called the Coast Guard repeatedly for help: "Mayday, mayday, mayday, Coast Guard—bridge down." The Florida Coast Guard and the Department of Transport were on the scene within the hour to help pick up survivors and to stop all traffic on the bridge.

In all 35 people plunged to their deaths that day, and only one person survived the tragedy. Wes MacIntyre, a vehicle maintenance man, was traveling over the bridge in the driving rain at 7.34 a.m. when suddenly his blue Ford Courier truck started to bob up and down and roll from side to side. He thought it was the wind and ignored it, but a few moments later he looked ahead and saw a ship in the water below and no bridge at all. His truck was airborne by the time he applied the brakes, and was dropping into the water. It bounced off the bow of the *Summit Venture* before sinking 30 feet to the seabed.

ABOVE: The cable stay main span of the new Sunshine Skyway bridge, which opened to traffic in 1987.

Wes MacIntyre returned to consciousness a moment later to find himself in his driving seat with the windows closed and water streaming into the cab of the truck. He took a deep breath, then forced the cab door open and swam to the surface. A crewman on the *Summit Venture* spotted him in the water and pulled him to safety.

In January, 1981, after much public debate and detailed investigation as to whether to repair or rebuild the original span, or construct an entirely new bridge, Governor Bob Graham announced the decision that an entirely new Sunshine Skyway would be built. It was to become one of the safest and most celebrated of modern bridges, and the longest cable-stay bridge in the USA.

The second Sunshine Skyway A bridge connecting St Petersburg and Clearwater across Tampa Bay has existed since 1954. The first bridge was built with an 846-foot, steel-truss-girder main span over the navigation channel. In 1971 an identical sister bridge was built alongside the old one, to increase the traffic flow. The original bridge was called the Sunshine Skyway in 1954 but, after the second bridge was added, it was officially renamed the Bill Dean Bridge after the chief of bridge design in Florida state. However, the public and press never really liked that name, preferring the magic of the original Sunshine Skyway.

ABOVE AND LEFT: A motorist view of the curved main span and the central cable stays.

Following the tragedy of the *Summit Venture*, the original bridges have now been replaced by an entirely new bridge running on a separate alignment, across four miles of Tampa Bay. The high-level approaches and the cable-stay navigation span were designed by the consultants Figg and Muller under their chief engineer, Jean Muller. The low-level approach spans were designed by the Florida Department of Transportation, while Parsons Brinckerhoff designed the bridge protection system. The dolphin bumpers and concrete islands placed around the main pier supports can take the impact of an 87,000-ton tanker traveling at 10 knots and not budge.

The cable-stay span is 1,200 feet in length, giving the roadway a maximum clearance above the water line of 193 feet. It contains 2.36 million feet of cable strand, which needed 2,500 gallons of bright-yellow paint to coat all the cable covers. The 70-foot-diameter pier support extends 175 feet from the water line to meet the bridge deck, and each one contains 13,000 cubic yards of concrete. The concrete pylon above the road deck tapers to a 50-foot diameter and finishes 431 feet above the water line. Each pylon supports two sets of 21 cable stays, picking up the bridge deck on each side of the span.

Construction work started in June, 1982, and the bridge was opened to traffic in April 1987 at an estimated cost of $245 million. In total the bridge contains enough steel to build a fleet of 746 Greyhound buses and enough concrete to form a 4-foot pathway from Pensacola to Key West. The 40-foot-wide, four-lane carriageway can accommodate 20,000 cars per day.

ABOVE: Sunset silhouette of the Sunshine Skyway.

Because the climate is subtropical, hurricanes can occur in the Tampa Bay area. So the bridge has been designed to withstand a wind speed of 240 m.p.h. and a gust of 290 m.p.h. The highest wind speed ever recorded so far in the Gulf of Mexico, into which Tampa Bay flows, was produced by Hurricane Camille, with a wind speed of 190 mph in 1969.

Like many bridges in the world, the Sunshine Skyway holds a superstition. Some fishermen who fish under the approach spans of the bridge believe that the body of a construction worker lies buried in the concrete in one of them—a victim of a construction accident. Some late-night motorists claim to have seen a woman on the main span waving frantically for help, but just as they slow down to stop the apparition is gone.

Pont de Normandie, Honfleur, France (1995) Another giant leap for progress in bridge engineering was taken when the Pont de Normandie was opened to traffic in 1994. We need to go back to 1931 when the George Washington Bridge smashed the longest-span record by over 60 percent, to find a jump to compare to the cable-stay span of the Pont du Normandie. The Pont de Normandie, which has a main span of 2,800 feet, has beaten the record held by the Skarnsundet Bridge of 1,740 feet set in 1991, and the Quingzhou Minjang Bridge built in China in 1996, with a span of 1,980 feet, by a big margin.

The geology of the site and the river bed and the light traffic flows the bridge was to carry were the two critical factors that tipped the balance in favor of a cable-stay bridge rather than a suspension bridge. The third factor, hardly ever admitted on record, was the emotive one of nationalistic pride in wanting to build the world's longest cable-stay bridge! The solution of a suspension bridge was ruled out because enormous anchorages would have to be built into the soft alluvium, since the site was devoid of good natural ground support.

From end to end the bridge is a mile and a quarter long, with extensive approach viaducts on both ends of the central span. The main span section under each tower, and the side span connecting the viaduct on both sides of the bridge, were designed as prestressed-concrete box sections, and built using the balanced cantilever method. The side spans and the main span section under the towers provide the rigidity for controlling the deformation of the large central span. Steel box girder rather than concrete was preferred for the 2,050-foot central span section, to reduce the dead weight of the span.

On July 5, 1994, the last of the steel box-girder sections weighing 200 tons was slowly lifted above the water of the Seine. The 32nd segment, which measures 75 feet wide, 16 feet deep, and 65 feet long, had taken three hours to move the 165 feet from the water level to the deck. Now all that remained to be

ABOVE: Temporary cables are attached to help support the main deck structure.

BELOW: Elevation drawing of the Pont de Normandie.

FOLLOWING SPREAD: Pont de Normandie is now as much a part of Normandy heritage as Calvados and Camembert.

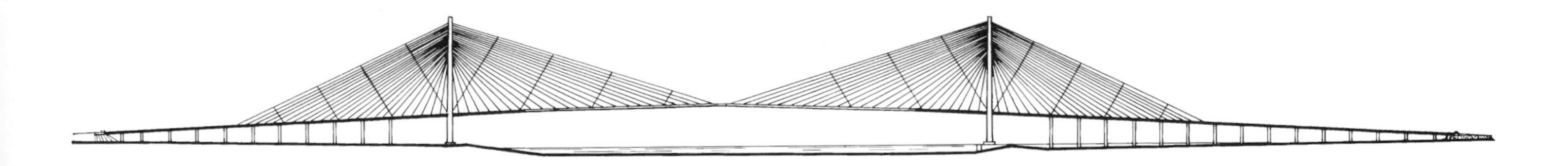

ABOVE LEFT: The cable stays stretch down from the pylon and are then fixed to the bridge deck.

ABOVE RIGHT: Looking up at one of the main pylons, and the web of permanent and temporary cable stays that sweep down to the deck.

done was to weld the box girder to the adjacent segment and then to join the two huge cantilever halves of the bridge, by inserting a small 8-foot-long keying section. The Seine can now be crossed near Le Havre without making a 30-mile detour to cross at the Tarcanville Bridge. It had taken four years of financial planning and technical study and six years of construction to complete the bridge. The two A-frame concrete cable towers standing guard over the estuary are as tall as the Montparnasse Building in Paris, France's tallest building, while the main span—which is held up by 184 gigantic stays—is as long as the Champs Elysées.

The stay cables were installed at the rate of two every day, with the geometry of the entire structure having to be adjusted constantly, as they were fitted. During the progress of construction computer specialists had to integrate and process up to 36 topographical measurements taken every day and assimilate more than 5,000 plans and sets of calculations to check the movement and stress effects on specific parts of the structure. Without the fantastic progress that has taken place with information technology and computer science in the past decade, this bridge would never have seen the light of day.

In windy squalls the construction engineers found that the longest cables began to vibrate at the same rhythm as the bridge deck, setting up dangerous harmonics. There was a risk that the whole bridge would begin to resonate like a gigantic harp. To prevent this happening, 32 bundles of transverse wire ropes were installed across the cable stays to dampen these oscillations. Stay cables can slacken under their own weight, just like a washing line. When the wind blows they tend to vibrate. If they are tied to one another, the livelier ones are calmed by the others.

Before the bridge was officially opened on January 20, 1995, it was given some punishing tests. First a fleet of 80 35-ton lorries maneuvered on the deck so that the vertical sag, the horizontal movement of the towers, the distortion, and load on the stay cables could be measured. The results tallied almost exactly with the engineered calculations. Then, a few days before the opening, a powerful oceangoing tug was deliberately moored to the deck, and then pulled on the mooring ropes before suddenly releasing its 140-ton strain, but the bridge deck did not waver.

"To send more than 70,000 cubic meters of concrete and 19,000 tons of steel arching into the sky between Honfleur and Le Havre would have been unimaginable twenty years ago," says Michel Virlogeux, lead designer of the Pont de Normandie. "This project has been a real scientific adventure from start to finish." It may take till the year 2001 and beyond to recover the $456 million that has been invested in the construction of this great bridge, not that the people of Normandy will mind, because they love the bridge. It has already become part of their heritage, like Camembert and Calvados.

ABOVE (CLOCKWISE): The anchor plate that connects the cable stays to the pylon head; Attaching the transverse cable ties to dampen any cable vibration in high winds; Abseilers checking the cable cover shields.

"This project has been a real scientific adventure from start to finish," says Michel Virlogeux, chief design engineer of the Pont de Normandie.

50
50
SMITH

A chapter of disasters

No history of bridges could be complete without knowing something about the many bridge collapses, loss of life, and the catastrophic failures that have darkened the horizons of bridge-building progress over the centuries. The most dangerous period for a bridge is when it is being built: temporary supports can fail, while human error can lead to neglect of a critical operating procedure or an underestimate of the stresses on the bridge. Occasionally forces of nature can be just too overpowering and overwhelming for a structure to resist. And just occasionally a bridge can be damaged by accidental impact from a ship or the derailment of a moving train. These are unforeseen events that the bridge designer would not have allowed for.

Many people may be under the illusion that bridge failures do not happen nowadays, but nothing could be farther from the truth. Nineteen ninety-eight was one of the biggest boom years for bridge building in the world, with some of the longest, tallest, heaviest bridges the world has ever seen being built.

It was also a bad year for bridge accidents. During 1998 there was a report almost every month of an accident or collapse of a bridge somewhere in the world—and there were fatalities.

ABOVE: Milford Haven, Wales—steel box girder span buckles during construction in June 1970.

OPPOSITE: Collapse of Cypress Viaduct, after the San Francisco earthquake of 1989.

LAUNCH SYSTEM SCRAPPED
after Japanese bridge fall

KURUSHIMA BRIDGE, JAPAN.

Japanese bridge engineers are conducting urgent safety checks on their most widely used incremental deck launching technique following the release of a scathing report on the country's worst bridge accident in decades, which killed seven workers. The accident occurred in June on one of the world's largest bridge projects, the £5.5 billion [approximately $8.8 billion] Kurushima crossing, near Imabari, southern Japan. Sections of temporary steel platform supporting an already launched side span were being dismantled and lowered on cables. Three cables gave way, tipping the 50 ton platform section and plunging workers on it 60m [196 feet] to the ground.

(from *New Civil Engineer*, September, 1998)

Injaka bridge collapse—

South African experts probe incremental bridge launch

Fourteen people died and thirteen were seriously injured when two of the Injaka Bridge spans collapsed during construction. The accident happened when a 27m [88 feet] long steel nose girder attached to the front of the first deck segment had just been pushed on to a temporary bearing on the second pier when suddenly both spans collapsed, dropping workers and a party of visitors 30m [98 feet] to the ground. A full scale investigation into tourteen people died and thirteen were seriously injured when two of the Injaka Bridge disaster has been launched by the South African Department of Labor, backed by the Police and Health and Safety Inspectorate.

The bridge collapse was all the more tragic as a party of guests invited to the site by the consultants VKE Engineers who designed the bridge, were injured, some fatally. What was intended to be a special occasion to celebrate the completion of the second span turned in moments into a scene of destruction. Among the dead was the bridge's designer 27 year old Marlieze Gouws, described by colleagues as a competent young engineer who had gained an honors degree with distinction.

(from *New Civil Engineer*, July, 1998)

BRIDGE COLLAPSE

(from *Civil Engineering International*, January, 1998)

One man was killed and two men injured when a temporary bridge collapsed during demolition near the town of Iaeger, West Virginia. Director of West Virginia highways construction Bob Tinney said the 50m [165 feet] long central span fell as contractor Battleridge Companies was preparing to lift it out of position. "The contractor had a crane tied to the span and I understand that connecting bolts were being taken out at the time of the accident," said Tinney. He added that the company had worked regularly for the state and had a good safety record.

Let's now go back in time to record the events of some of the worst bridge disasters in the history of bridge building.

Wheeling Bridge, Ohio, 1854 The collapse in 1854 of the 1,000-foot-span Wheeling Suspension Bridge, the longest bridge in the world, was a terrible lesson that was lost to the profession, because history was to repeat itself almost 100 years later. David Steinman recounts in his book *Bridges and Their Builders* a remarkable eyewitness account of the Wheeling collapse, which was reported in the Wheeling *Intelligencer*:

ABOVE: Wheeling Bridge, Ohio, USA, after it was rebuilt.

With feelings of unutterable sorrow, we announce that the noble and world renowned structure, the Wheeling Suspension Bridge, has been swept from its stronghold by a terrific storm ... At about 3 o'clock yesterday we walked toward the Suspension bridge and went upon it, as we frequently have done, enjoying the cool breeze and the undulating motion of the bridge ... We had been off the [bridge] only two minutes and were on Main Street when we saw persons running toward the river bank: we followed just in time to see the whole structure heaving and dashing with tremendous force.

For a few moments we watched it with breathless anxiety, lunging like a ship in a storm; at one time it rose to nearly the height of the tower, then fell, twisted and writhed, and was dashed almost bottom upward. At last there seemed to be a determined twist along the entire span, with about one half of the [bridge deck] being reversed, and down went the immense structure from its dizzy height to the stream below, with an appalling crash and roar. Charles Ellet completed his greatest work, the Wheeling Suspension Bridge, over the Ohio River in 1848. Six years later, on May 17, 1854, the great bridge was destroyed by the wind. The bridge that Ellet had designed was quite capable of supporting its own weight and the loading from bridge traffic of horses, carts, wagons, people, and cattle. It was robust enough to resist the force of a considerable wind, but not the vibration patterns set up by the wind on the deck, which in turn could cause the cables to sway. The Wheeling Bridge illustrated how little the principles of aerodynamic stability were understood during this period, with the exception of John Roebling. He had worked out why the Wheeling Bridge had failed and seemed to have anticipated such problems in his designs, introducing traverse or cable stays between the bridge deck and the suspension cables.

No one was killed or injured. The bridge was later rebuilt by Charles Ellet with transverse stays and a stiffer bridge deck structure, and reopened in 1860.

Ashtabula, Ohio, USA 1876 In America in the nineteenth century no railroad promoter could afford to build bridges entirely in cast iron, however strong and efficient it could be. A railroad could reap handsome profits by building a rail track across hundreds of miles of the countryside just adequate for carrying a steam locomotive pulling a few light rail cars. A rail bridge had to be built at the smallest cost, capable of supporting the loads of a slow-moving train and nothing more. An iron arch, while being cheaper than a tubular-iron truss, was still more expensive than a composite wood and iron truss. So the wood and iron truss was what the railroad settled for. Many of their bridges were based on the Howe truss—it was one of the more popular designs. But, as the trains became heavier and the locomotive more powerful in the 1850s, many Howe-truss bridges began to collapse.

In 1865 the Lake Shore & Southern Michigan Railroad faced the problem of replacing an important bridge at Ashtabula, Ohio, over Ashtabula Creek. The creek was shallow but it ran through a gorge 700 feet wide and 75 deep. A new bridge was built, made entirely of wrought iron in the form of a modified Howe truss. It represented a new approach to rail-bridge construction in the USA. It had a span of only 150 feet because the railroad company had built up the embankment on each side of the gorge.

ABOVE: Ashtabula Bridge in service before it collapsed.

Eleven years passed without incident until the night of December 29, 1876. The Pacific Express, a westbound 11-car train with double locomotives, had started to travel across the Ashtabula Bridge in a blinding snowstorm. Halfway across, Dan McGuire, the driver of the front locomotive, felt a huge shudder and the sensation of the train going uphill. He reacted by immediately opening the throttle to make the locomotive surge forward and, as it did, he heard a loud grinding sound. It was the tender of his locomotive scraping against the bridge abutment.

In the next moment there was a mighty crash as the second locomotive smashed into the same abutment. McGuire's locomotive then sped forward along the track and off the bridge as the coupling bar connecting the tender to the second locomotive was sheared. He brought the locomotive to a halt and then backed it along the track. By the time he jumped off and had run to the edge of the embankment, a ghastly illumination was glowing through the snowstorm, down in the gorge. The rest of the train, composed of two express cars, two baggage cars, two passenger coaches, a smoking car, a drawing room car, and three sleeper cars, had fallen into the gorge, crashing into one another. The wood-burning stoves in the cars had set light to their wooden hulks and were ablaze.

Of the 90 people killed, most of them died instantly from the crash as the cars plunged into the gorge. Those who were not crushed to death were uncertain whether to remain inside the cars and risk being burned, or face being frozen in the icy waters of the Ashtabula as they got out. Out of the coaches, hands and arms were thrust forward and slowly a trickle of confused and terrified passengers clambered out and staggered up the embankment free of the fires. It took a while for help to arrive.

This collapse of a wrought-iron bridge reverberated throughout the US. Newspapers and magazines carried stories about the collapse: the *Iron Age* magazine voiced the fears of the people, saying, “We know there are plenty of cheap, badly built bridges, which the engineers are watching with anxious fears, and which, to all appearances, only stand by the grace of God!” While the *Nation* summed up the apathy for better safety in design declaring, “By such disasters and by shipwrecks are lives in these days sacrificed by the score, and yet except through the clumsy machinery of the coroners’ jury, hardly anywhere in America is there the slightest provision made for an inquiry into them.”

The coroner put the blame of the Ashtabula disaster down to the designer of the modified Howe truss, Amasa Stone, and on Charles Collins, the chief engineer of the rail line, who survived the disaster by being in the last coach. Collins was unfairly blamed and was made a scapegoat by the press. He was a sensitive, gentle person, who suffered horribly with personal agonies of guilt and finally committed suicide. On the other hand Amasa Stone, a more ebullient, arrogant individual, defended the design of the bridge and his integrity very robustly.

The Ashtabula disaster was down to the lack of knowledge of cast-iron behavior—its low resistance to tensile forces, the need for additional diagonal

BELOW: A chart of the collapsed bottom chord of the truss, where it fell in Ashtabula Creek, as observed about three weeks after the disaster took place.

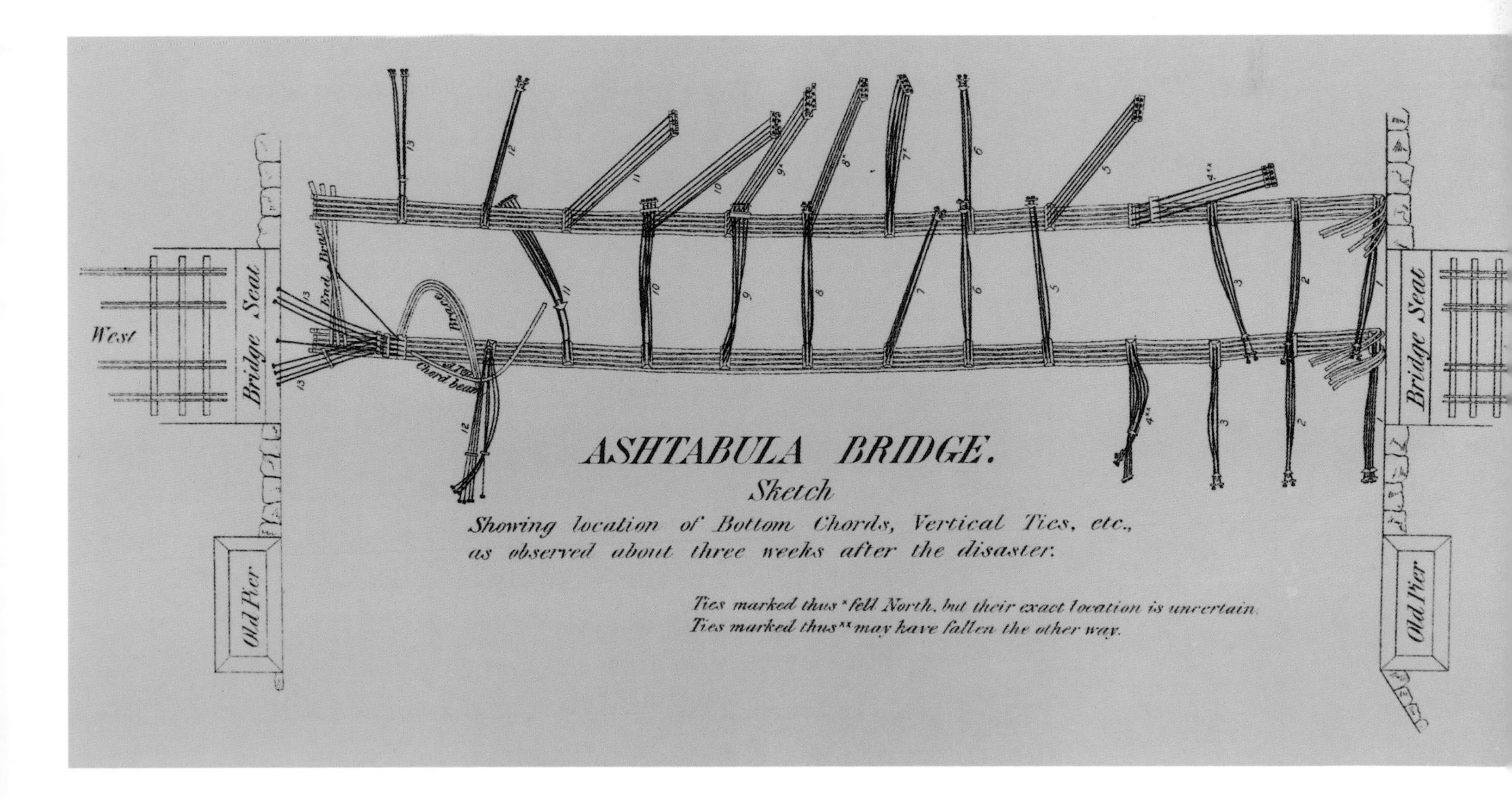

bracing of trusses, the brittle failure of wrought iron under repeating tensile loading—and the cavalier fashion by which the railroad cut corners and often disregarded the design capacity of a bridge. It was the derailment of the train, however, in the heavy snowfall on the track, that shifted most of the weight of the train to one side of the truss, causing the stresses to be reversed and to go into tension, which then caused the structure to buckle and collapse.

There were many truss bridges built in the US that were just like the Ashtabula. Two years later a wood-and-iron Howe truss at Tarifville, Connecticut, fell under an excursion train, killing 17 people. There was no derailment: the bridge simply collapsed. During the 1870s no fewer than 40 bridges a year fell, half of them purely timber structures. But the iron-bridge failures made the headlines, because they were the biggest and their collapse resulted in the greatest loss of lives. In the ten years after the Ashtabula tragedy, 200 bridges collapsed, many involving major loss of life.

It marked the end of the cast-iron bridge era in the US and the introduction of major reforms in bridge safety standards, of a state bridge inspectorate, and the requirement by law that a bridge design specification should be passed by the ASCE, the American Society of Civil Engineers.

The Tay Bridge disaster, 1878, Dundee, Scotland Two years after Ashtabula, and almost exactly to the day, a worse disaster shocked the bridge world and dented the pride of a great bridge-building empire.

Thomas Bouch, the designer of the completed Tay Bridge, was busy designing the biggest bridge in the world, the Firth of Forth, when he received the awful news "that the bridge was doon." Thomas Bouch was until that moment the most renowned of living engineers in Britain. In June that year he had been honored when Queen Victoria and her royal train stopped for a while at Tay Bridge Station to receive an address of welcome and to meet with leading officials, including Bouch. The Queen's train had made a special detour from Balmoral on its way down to London to cross the newly opened Tay Bridge. Shortly afterwards Bouch received his knighthood from Queen Victoria, standing alongside Henry Bessemer, who was the inventor of a process for converting iron into steel, at Windsor Castle.

Although the Tay Bridge was an extraordinarily long bridge and the biggest bridge in overall length in the world, it had a fairly standard design, and did not feature any new ideas or construction techniques. Each big lattice girder in wrought iron was riveted together on the south shore, then floated out on heavy barges and raised onto the piers by hydraulic jacks. Several of the foundations were sunk by pneumatic caisson.

The completed bridge was a thin, long ribbon reaching out from the shore on slender supports rather like a fragile seaside pier, with the distinctive high girder spans in the middle to provide clearance for shipping. There was nothing spectacular or inspiring about the bridge and there were no adverse comments about its safety, with the exception of a chance remark made by Major General Hutchinson of the Royal Engineers, when inspecting and testing the bridge to issue its structural worthiness certificate on behalf of the Board of Trade. In his report Hutchinson made some minor criticisms of the bridge and recommended that trains be restricted to a maximum speed of 25 m.p.h. At the end of the report he made a casual reference to the wind: "When again visiting the spot, I should wish if possible to have an opportunity of observing the effects of high winds when a train of carriages is running over the bridge." This observation was never made in the final inspection as Hutchinson fell ill shortly afterward and his replacement officer did not attach any importance to it.

It was already dark when the storm of Sunday, December 28, 1878, struck Dundee. The strength of the storm increased over the next few hours into a gale as the wind speed climbed from Force 10 to Force 11 approaching 100 m.p.h. The gale blew across the open stretch of the Firth of the Tay, buffeting the bridge. It was just gone seven o'clock in the evening when the mail train from Burnt Island arrived at Tayport and slowed down to three miles an hour to be given the signal to cross the bridge. Thomas Barclay, the signalman, handed the fireman the clearance baton, and then hastened back to his hut to send a wire signal to the north shore to let them know the train was on its way.

From his hut Thomas and a colleague, John Watt, watched the train's lights as it made its way over the bridge in the driving rain and the howling gale. There was a spray of sparks from the wheels and then three flashes and one big flash. Suddenly Watt could not see the taillights of the train.

BELOW: Illustration of the Tay Bridge shortly after it was built.

"There's something wrong with the train," said Watt. Barclay was not so sure and thought the train was hidden in the curve of the bridge and was temporarily blocked from view. After a while, when no lights reappeared on the far side, Barclay immediately rang the bell to signal the north shore, but there was no answering bell. He tried his speaking tube but got no reply.

Thomas and Watt then ran out of the hut into the gale and started to scramble across the bridge. They had to get on their hands and knees for fear of being blown into the estuary, but after 20 yards they gave up and turned back. They climbed off the bridge down onto the bank at the water's edge in the rain-sodden blackness in the hope that they might see something. Incredibly, for a brief moment, the moon came out and to their horror they saw that the middle section of the bridge where the 13 high girder spans should have been had disappeared.

The two men turned and scrambled up the banks to raise the alarm that the "bridge was doon." A tense crowd had gathered in the harbor on the south shore, aware that a disaster had happened because they too had seen sparks, flashes, and columns of white spray leaping up from the black waters. A ferryboat was ordered out but returned at midnight with no sign of any survivors despite carefully negotiating the pier stumps of the bridge.

Farther downstream there were reports of mailbags being washed ashore. The news was telegraphed to London and the next day the whole world was in shock. Seventy-five people died in the disaster but only 46 bodies were ever recovered; 29 have remained buried in the Tay to this day.

BELOW: Looking for survivors in the icy waters of the Tay.

What caused the disaster? Two points of view emerged immediately. The religious extremists blamed the railroad for running trains on Sundays, saying it was the "hand of God" determined to guard the Sabbath. The business community and many others thought the man to blame was the designer.

An investigation was conducted by a court of experts appointed by the Board of Trade, and commenced on the spot in Dundee, as divers searched the Tay for bodies. They interrogated dozens of witnesses, listened to the facts, the explanations, and the theories of what may have caused the collapse from everyone including Sir Thomas Bouch. Several damning things emerged concerning the quality of the cast iron in the girders and the piers, which had been found with "Beaumont Eggs." This is the practice of filling holes found in the cast iron with a mixture of beeswax and iron filings. It turned out the inspector responsible for the maintenance of the bridge was inexperienced and incompetent.

ABOVE: What remained of the support legs and the high girders after the Tay Bridge collapsed on Dec 28, 1878.

Locomotives were often allowed to race across the bridge exceeding 25 m.p.h. to overtake the ferry. But the most appalling discovery was when they found out that Sir Thomas Bouch had scarcely considered the full effects of the wind in the Tay when designing the high girders of the bridge. Bouch had allowed for a wind pressure of 20 pounds per square foot acting on the bridge. His calculations were based on wind tables prepared by John Smeaton in 1759, about 120 years before.

Smeaton's tables were not inaccurate: they were simply not appropriate for an open estuary like the Tay, where wind gusts would greatly exceed the steady uniform pressure assumed by Smeaton. Bouch had also consulted with the Astronomer Royal, Sir George Airy, for advice and he confirmed that "the greatest wind pressure to which the plane surface like that of a bridge would be subject to is ten pounds per square foot."

In summing up Henry Rothery, Her Majesty's Wreck Commissioner and one of the members of the three-man court, stated in his report, "I do not understand [why] my colleagues differ from me in thinking that the chief blame for the casualty rests with Sir Thomas Bouch, but they consider it is not for us to say so ..." And he went on to explain his reasons: "Engineers in France made an allowance of fifty-five pounds per square foot for wind pressure and in the United States an allowance of fifty pounds was made."

Sir George Airey had written to the court and advised them that in his opinion a wind pressure of 120 pounds per square foot should be assumed for bridges, not ten pounds per square foot! Rothery then spelled out the blame: "The conclusion ... is that this bridge was badly designed, badly constructed, and badly maintained and that its downfall was due to inherent defects in the structure which must sooner or later have brought it down. Sir Thomas Bouch is, in our opinion, mainly to blame."

The bridge had been strained by previous gales and by the trains that ran on it at excessive speeds. The wind acting on the bridge during a gale is not a single pressure, but a series of gusts. These gusts would have a greater overturning effect on the bridge than the measured pressures of a Force 11 gale. The twin trusses and flooring of each high girder span would present a large exposure area, with the force from such gusts enough to possibly collapse the bridge, particularly as the span may not have had sufficient lateral stability, after being shaken by previous gales and the vibrations from passing trains.

It is likely that, as the train traveled over the juddering high girder spans and was slammed broadside by the high winds, the riveted connections of a girder and pier came loose and something broke. A few moments later the pier collapsed, and a girder fell, then girder after girder and pier after pier just crumpled like a deck of cards, plunging the locomotive, the passenger cars, and the people 88 feet down into the freezing waters of the Tay.

Sir Thomas Bouch was destroyed: he was dismissed as the engineer of the Forth Rail Bridge and died from pneumonia not long afterwards, a bankrupt and broken-hearted man. But was it really all his fault?

OPPOSITE ABOVE: The critical cantilever span of the Quebec Bridge being erected.

OPPOSITE BELOW: Tangled heap of the cantilever span lies strewn on the shore of the River Lawrence.

The Quebec Bridge disaster, 1907, Canada The arrival of the twentieth century heralded a new dawn for bridge construction. No longer was bridge building dominated by the masonry arch and the timber and wrought-iron truss. The arrival of steel provided a real battleground for bridge forms—on one side there was the suspension-bridge lobby championed by the mighty Brooklyn Bridge, and on the other the massive cantilever truss of the Forth Rail Bridge. Was the cantilever-truss arch a more stable structure than a suspension bridge? Which bridge form was quicker to erect, cheaper on cost, safer under wind loading, and more able to support a fast-moving railroad?

The American railroad companies preferred the cantilever-truss bridge to the suspension bridge because they felt its rigidity made it ideal for the heavy loading of the railroads. When a new rail bridge was proposed over the St Lawrence River valley in Quebec, Theodore Cooper, one of the most eminent bridge engineers, was asked to design a long-span cantilever bridge. It was going be the longest-span bridge in the world, beating the span of the Forth Rail Bridge by 100 feet.

However, the first estimate of the weight of steel for the bridge pointed to an enormous cost, and Cooper and his team were put under pressure by the rail board to use all their resourcefulness to keep down the tonnage. The design that

#24
9-7-07

ABOVE: 9000 tons of steel fell suddenly to the ground, when the span of the first bridge collapsed in 1907.

Cooper evolved, although inspired by the Forth Bridge cantilever, differed in construction concept because it had only one main span and two end spans. The Forth was built with the cantilever spans in balance, and built out equally from the central supports. The relatively short end spans at each end of the Quebec Bridge had to secure the longer length of cantilever main span during construction.

Work was going along well, the end spans had been built, and the south cantilever arm was nearing mid-span, when the resident engineer, Norman McClure, sent Cooper a telegram telling him that the cantilever was starting to deflect downward by a fraction of an inch. McClure sent more telegrams to Cooper on subsequent days, as the deflection increased, urging him to visit the site. Then on August 27 he sent a final telegram to Cooper saying "erection will not proceed until we hear from you and from Phoenixville." The Phoenix Bridge Company, who were erecting the bridge, were based in Phoenixville and that's whom McClure was referring to in the telegram. They had responsibility for the bridge works during its construction and naturally had to be consulted.

The next day Cooper replied, ordering an immediate investigation into the deflection, aware that he may have pared the structure down too drastically to save on the cost of steel. His message did not state that all work should be halted, but neither did it say carry on regardless. McClure decided it was best to stop all construction work that day and left for New York to see Cooper (who was 70) in person. A contractors' superintendent on site, not aware of any potential danger—one assumes he was not subject to directives by the resident engineer—sent the men to work on the cantilever arm as usual the next day, August 29. Even though the deflection had now increased visibly, a crane was moved out onto the span during the day.

On the evening of August 29, 1907, just before work stopped for the day, the sound of tearing metal filled the air, signaling the worst bridge construction disaster ever recorded. The incomplete cantilever arm of the south span broke away and 19,000 tons of steel crashed into the St Lawrence river with 86 men on it. Eleven men luckily survived the tragedy.

Insufficient stress-analysis theory, construction knowledge, and engineering mathematics—these were the real deficiencies with the bridge, not the fact that the steel had been cut right down to the bone. Some of the compression elements in the truss of the cantilever arm were subject to a large squashing force that tended to bend them outward, making them buckle and ultimately give way. The rupture point of the Quebec Bridge was at the connection to the main chord. The joint actually buckled at the connecting plate, which was not made rigid or strong enough. It had only two connecting rivets when it should have had eight, according to the inquiry team. In fact the joint had only 30 percent of the strength of the compression member it was connecting.

It was clear that the accepted design rules at the time, based on the empirical design of compression members and connecting plates of much smaller structures, was unsuitable for very large-scale structures. Theodore Cooper, one of the finest bridge engineers of his time, who had worked as Eade's right-hand man on the St Louis Bridge, died shortly after the event, a broken man. More scientific study was necessary to understand the behavior of buckling in compression members and the proper detailing of construction joints.

That said, ten years later in 1916, by which time the second cantilever Quebec Bridge had been redesigned with K-bracing and was nearly built, a second tragedy happened. The central suspended span was being jacked up into position from a floating barge, when one of the jacks failed with the steel truss 15 feet in the air. Five thousand tons of steel crumpled and fell into the St Lawrence, killing 11 workers.

Quebec Bridge, with two horrendous collapses costing 87 lives, had taken over ten years to finish, just to claim the world's longest span by 100 feet. Was it worth the price?

BELOW: The rebuilt Quebec Bridge, 1917 with the additional "K" bracing.

ABOVE: The final moments before "Galloping Gertie" plunges into the Pugent Sound.

RIGHT: A 600 ft section of the Tacoma Narrows bridge is torn away, as the structure writhes and twists uncontrollably.

The Tacoma Narrows Bridge failure, 1940, Pugent Sound, USA With the Golden Gate and Oakland Bay bridges recently completed, America led the world in suspension-bridge construction, and now embarked on a crusade for even more graceful, slender, and daring span designs. None was more elegant than the span of the Tacoma Narrows Bridge over the beautiful Pugent Sound in Washington State, which opened to traffic on July 1, 1940. From the very day it opened there was something extraordinary about it. Motorists crossing the 2,800-foot span frequently saw the car ahead sink into the road and then rise again. No one was alarmed, since the engineers assured drivers that the structure was safe, even though the span undulated in breezy conditions. Drivers began to really enjoy the novelty of the gentle roller-coaster ride across the flexible span of the bridge. It became quite an attraction and soon acquired the nickname of "Galloping Gertie." Toll revenues from the bridge ran higher than expected—everyone was pleased.

Four months later, on November 7, with a moderately high wind blowing at about 44 m.p.h., the span suddenly went into an alarming series of rolls and pitches. One side of the road dipped and the other side rose, as a ripple wave traveled the length of the span. The bridge was closed to traffic. One motorist stopped his car partway across as the deck motions became quite terrifying. He got out of the car but his dog would not leave it. After a few moments, he walked back alone. A few minutes later, some suspension hangers tore loose and a 600-foot length of concrete deck fell into the sound. The remainder of the deck now whipped and thrashed about in a frenzy, until it too ripped free of the hangers, sending the whole of it into the Sound. With it went the car and the dog.

LEFT: The second Tacoma Bridge was designed with a deep truss deck to prevent it from vibrating in high winds.

The Tacoma collapse was the most spectacular bridge failure of all time. Every moment of its dance of death was filmed and recorded by Professor Farquharson of the University of Washington. The film footage of the collapse became a newsreel classic. The designer of the Tacoma was blamed for its collapse. He was Leon Moissieff, who was responsible for the design of the beautiful Manhattan Suspension Bridge, and a leading consultant on the Golden Gate, Oakland Bay, and Bronx Whitestone bridges with Othmar Amman and many others. At 68, he was one of the most experienced suspension-bridge engineers in America. When he submitted his slim, plate-stiffened deck-girder span for the Tacoma Bridge, based on the success of the Cologne Bridge designed by Fritz Leonhardt, and which had been used by Amman on the Bronx Whitestone Bridge, nobody argued that it was too slender or unsafe in high winds. Quite the reverse was true—bridge engineers admired and raved about Moisseiff's streamlined design.

Steinman summed up the tragedy of the Tacoma poignantly: "The span failure is not to be blamed on Moissieff alone; the entire profession shares in the responsibility. It is simply that the profession had neglected to combine and apply in time, the knowledge of aerodynamics and dynamic vibrations with its rapidly advancing knowledge and development of structural design." The lessons of the Wheeling Bridge had been forgotten in the intellectual challenge of designing long-span structures during the 1930s and 1940s.

Thereafter, the suspension-bridge decks in the United States were designed with deep stiffening trusses to avoid dynamic instability in high winds. The Tacoma Bridge was rebuilt later with a deep truss deck, while the Golden Gate, the Bronx Whitestone, and many others were retrofitted with stiffening trusses.

West Gate, Australia and Milford Haven, Wales **Failure of steel box girder bridges in the 70s** In the 1970s the world's attention was focused on the collapse of four steel box girder bridges during construction. The bridges were in Vienna over the Danube, in Koblenz over the Rhine, in Milford Haven, Wales, where four people were killed, and in Melbourne over the Yarra River, which killed thirty-five people.

ABOVE: Milford Haven, Wales during the rebuilding work of the collapsed span.

It has become increasingly clear over the centuries as stronger materials are developed and new construction technologies evolve, that the limiting factors are not technological nor economic constraints, but the human ability to effectively communicate with one another and to decide on priorities. The more advanced the technology and the greater the economic pressure to save money, the more critical become the consequences of human actions or the lack of them. This was how bridge experts summed up the cause of failure of a series of steel box girders bridges in the 1970s, the worst of which was the West Gate Bridge collapse which occurred only four months after the Milford Haven collapse.

On October 15, 1970, the west cantilever section of the main span of the West Gate bridge suddenly crashed, bringing down 1,200 tons of steel on to workmens' huts below. Thirty-five people lost their lives, many more were seriously injured. How could such a collapse have happened only four months after a similar box girder span buckled and collapsed in Milford Haven, Wales? Both had been designed by the same team of consulting engineers.

One of the problems of the steel box girder is that during construction as the span cantilevers further out, the tension and compression stress in the beams is often greater than that from the eventual loads that they have to carry. As a result of this the deflection or sag at the tip of the cantilever may become very pronounced. Usually the ends of the cantilevers have to be hydraulically jacked up to align and connect the two cantilevers. This is a very time consuming operation. For the West Gate box girder, to make construction easier, each cantilever box section was prefabricated in two halves, and then hoisted into position and bolted together. A misalignment occurred between the two halves of the west cantilever section when it had reached 300 feet. It was decided to ballast one half of the deck with 60 tons of concrete to deflect it sufficiently to join it to the other half, rather than to jack up the span—to save time. A buckle appeared in the central plate that joined the two halves of the deck. The cause of the central plate buckle was being investigated when the 376-foot cantilever fell to the ground, without warning.

The international enquiry team into the collapse, organized by the British Government, was also given the responsibility for establishing the rules for

ABOVE: West Gate Bridge seen here with one of the box girder spans completed.

LEFT: 1200 tons of steel buckle and fall 300 ft to the ground on October 15, 1970.

ABOVE: The rebuilt West Gate Bridge over the Yarra river in Melbourne.

appraising steel box girder structures during construction, and for setting standards of safety checks and site procedures. In Britain traffic was restricted on all steel box girder bridges, pending a full structural investigation on each bridge, even though quite a few of them had been opened for several years. Mr Justice Barber of the New South Wales Supreme Court, who headed the enquiry team, concluded in his report:

> *"While we have found it necessary to make some criticism of all the other parties, justice to them requires us to state unequivocally that the greater part of the blame be attributed to (the consultants) Freeman Fox and Partners."*

The story of the West Gate collapse is about the complex relationship on site between the designer, the contractor, and specialist teams, and the lack of adequate standards and routines for checking and approving the prefabricated sections of the box girder during the construction.

Earthquake damage in the 1990s Volcanic eruptions apart, there is nothing on earth more frightening or devastating than an earthquake. In impact it is the equivalent of shock waves below ground of several atomic bomb blasts. Below the sea, an earthquake can give rise to tidal waves as high as 200 feet, which can submerge coral islands and wreak havoc with ships, harbors, and buildings some distance inland.

In recent years there have been significant earthquakes in Kobe, Japan, San Francisco, and Los Angeles that have collapsed bridges as well as buildings. The alarming concern for bridge engineers was that, having designed the bridge supports to resist the forces of an earthquake of a certain magnitude, they were shocked when many of their bridges collapsed. Assumptions about the most effective ways to absorb or resist the earthquake forces are still being studied and analyzed. No one can be certain that every eventuality will be catered for, but engineers in both Japan and California have come up with a new bridge-wrapping protection to the support columns and piers. They hope this will give the support added resistance against earthquake vibrations, but no one is absolutely sure.

It is a sad reality that it is only when something has been tested to the limit, and failed, that we can begin to fully understand the limits of the integrity of its construction, the behavior of the material it was built with, and the rigor of the structural analysis that the design was based on. With progress and change in our society continuing to evolve so rapidly, we may expect to stumble again on occasion.

ABOVE: The collapsed top carriageway of a double deck viaduct in San Francisco after the earthquake of 1989.

LEFT: "There is not much a bridge engineer can do to make a design completely safe against the devastating intensity of a big earthquake." San Francisco 1989.

6

Bridges and men

We often ask the same question about great ships, supersonic aircraft, classic cars, tall buildings, and bridges: Who built them? Sometimes we are given a clue to the identity of the designer by the name given to these artificial wonders—for example Bugatti, De Havilland, Rolls-Royce—but quite often they are named after the company that owns it, the city where it was built, or the river it crosses.

In the history of bridge building there have been a number of outstandingly brilliant designers, individuals who almost single-handedly changed the way bridges can be built to span farther, using daring and innovative concepts and new materials and assembly methods. Whether a bridge was built in a period when only stone and timber were available or a time when cast iron and wrought iron had been developed but not steel, or in an era when there were no computers or electronic measuring devices, just simple hand calculations, the rules of geometry, and the tape measure; certain individuals possessed a talent and skill far beyond those of their contemporaries. But their names do not appear on a bridge nor have any bridges been named after them. So who are they?

In this chapter the story of the great bridge engineers is told in the spirit of the prologue to David Steinman's book *Bridges and Their Builders* as a "heart-stirring

ABOVE: A precast segmental box girder section being lifted into position on one of the Florida Keys bridges.

OPPOSITE: The pedestrian walkway on the Brooklyn Bridge in New York is a popular place for tourists in summer.

narrative of high adventure and dramatic interest ... an epic vision of courage, hope and disappointment ... [for] a bridge is more than a thing of steel and stone: it is the embodiment of the effort of the human [mind], the heart and the hands."

Before beginning this tale of human endeavor let's just mention a few great men whose stories have not been told here because of space restrictions. Let's pay tribute to the past masters—the Roman engineer Caius Julius Lacer, builder of the Puenta Alcántara in Spain; to the leader of the French "Brothers of the Bridge" in the Middle Ages, Frère Benoit, a.k.a. St Bénézet, for building Pont d'Avignon; the English chaplain Peter Colechurch, who also belonged to a "Brothers of the Bridge" order, for building Old London Bridge; Taddeo Gaddi for the Ponte Vecchio; Antioni Da Ponte for the Rialto Bridge; Ammannati for the Santa Trinità; John Rennie for the New London Bridge; Isambard Kingdom Brunel for the Royal Albert Bridge at Saltash; Robert Stephenson for the Britannia Bridge; Gustav Eiffel for the Garabit Viaduct; James Eades for the St Louis Bridge; Robert Maillart for his concrete artistry and the Salgina Gorge Bridge; Gustav Lindenthal for the stunning steel arch of the Hell Gate Bridge; David Steinman for the Mackinac, the beautiful St John's and the Henry Hudson bridges; and Charles McCullogh for his "art deco" style of bridges.

And in recent times those on the short list toward immortality would include the bridge engineers Christian Menn, for his great teaching skill and aesthetic design of bridges; Jorg Schlaich for pioneering designs in cable-suspension bridges; the mercurial Santiago Calatrava for showing the world that modern bridge design can be an art form; Jiri Strasky for inventing the stressed-ribbon bridge; Michel Virlogeux for stretching the limits of the cable-stay bridge with the Pont de Normandie; and the bridge architects Alain Spielmann and Chris Wilkinson for putting architecture back into bridge design.

Now we will look in more detail at some of the great men who have populated the world of the bridge.

Jean-Rodolphe Perronet (1708–96) Born in Suresnes outside Paris, Perronet died in Paris at the ripe old age of 88, in the happiest circumstances he could imagine, while supervising the completion of the Pont de la Concorde over the Seine, during the French Revolution. Perronet was the son of an army officer and when he was six his parents took him to the Tuileries, the French royal residence adjacent to the Louvre. He enjoyed a lucky childhood, for in Paris the young prince who was to become Louis XV was being entertained in an adjoining garden. The prince saw Perronet and asked him to come and play. Out of this chance meeting a lifelong friendship was born, which brought Perronet many special privileges and personal favors.

When young Perronet had grown into manhood he asked to follow in his father's footsteps, but rather than be an officer he wanted to train as a military engineer. No doubt he had been inspired by the great military engineers of the day, among them Jacques Gabriel, who had just founded the first government engineering department for the scientific advancement of bridge building—the Corps des Ponts et Chaussées. Plans of all roads, canals, and bridge works in central France had to be approved by this influential and prestigious body of engineers, who were all graduates of the Ecole de Paris.

However, Perronet's luck was to run out, since he was not admitted to the elite military engineering college; because he was not one of the three candidates selected of the many applications that were received that year. So he studied architecture instead at the Ecole de Paris and later started on a career building bridges. He studied trigonometry, algebra, architectural history, and mechanical science, using multiplication tables as his computer in order to work out the arithmetic of mass, area, and force.

ABOVE: Jean Rodolphe Perronet

Good drafting skills were essential in his profession and many hours were spent perfecting neatness and clarity of line drawings using the quill pen and leaded pencil, aided by only by a ruler and compass. By night he must have worked under the light of flickering oil lamps, sitting studiously hunched over a crude wooden table, wearing gloves and warm clothes in winter to keep out the cold.

During the first half of the eighteenth century in France, as Perronet was gaining experience as a bridge engineer, a number of important stone-arch bridges were built on the Seine and Loire rivers. Perronet assisted Hapeau, the chief engineer of the Corps des Ponts et Chaussées, whom Louis XV had appointed to design and supervise the construction of a new bridge at Blois when the famous old one was washed away in a flood. After Hapeau's death, Perronet supervised the completion of bridges that Hapeau had started over the Loire at Orleans and over the Seine at Mantes. He was an able administrator, a brilliant organizer and motivator of men, who was hardworking and earnest, with a work ethic that he expected others 6 follow.

It was while working on the Mantes Bridge that Perronet made the first of his momentous observations. When the first stone arch was nearly complete and the second was getting under way, he noticed that the pier between them was leaning slightly toward the unfinished arch. Many arches with a pier-width-to-span ratio of 1:5, like those at Mantes, stand quite safely when the bridge is completed, even if the piers probably swayed a bit during construction. This temporary condition was not considered a problem. But Perronet's curiosity as to why the piers should lean led him to conclude that the whole group of arches must provide support to one another when the bridge was complete, and that the thrust from each arch—which had caused the piers to lean during construction—was transferred to the abutments at each end.

What Perronet discovered was that the arches of these spans were not truly independent of each other. He was the first man to discover that the horizontal

ABOVE: Typical wooden framework for centering the stone arch. A drawing prepared by the office of Perronet.

thrust of the arches was carried through the arch spans and that the piers carry only the vertical load and the difference between the thrusts of the adjacent arch spans. By keeping the arch span the same there would be no thrust on the piers, so the piers could be greatly reduced in thickness. But the big problem was maintaining the stability of the pier during construction. Perronet's ingenious solution was to have all the arches in place before the centering was removed, and to build the arches simultaneously, working from the piers toward the crown to minimize any thrust on the piers.

With this discovery, there were suddenly two great advantages in building stone-arch bridges. The more slender piers would widen the navigable waterway and present less of an obstruction in the river for scour damage to the foundations. The arch span could be made flatter, by transferring the arch thrust back to the abutments, and in raising the haunches the arch could clear the waterway as much as possible. Perronet's ratio of pier width to arch span was about 1:10 and 1:12, compared with the customary 1:5 ratio in his day. And, instead of the elliptical three-segmental arch that Hapeau and Gabriel before him had designed, Perronet designed the arch as a segment of a large circle. It was an arch that was esthetically more pleasing than the changing radius of the segmental arch, prompting him to remark, "Some engineers, finding that the arches … do not rise enough near their springing, have given a larger number of degrees and a larger radius to this part of the curve … [Such] curves have a fault disagreeable to the eye."

The first bridge that Perronet designed was a single-span arch at Nogent on the upper Seine and, although he could not put all his theories to the test, the elliptical-arch span was a radical departure from other arches. A few years later he was asked to build a stone-arch bridge over the Seine at Neuilly, to the west of Paris. The time had come to show the world one of the most revolutionary creations in stone-arch bridge designs.

But as usual Perronet had to defend his ideas. Many prominent bridge engineers thought that the bridge would never stand up, that the King's money was being wasted, and that people would be killed building the bridge and using it if it ever did stand up. Perronet's nerve held out. He explained the principle of his design and how all the centering would be in place while the arches were being built and why the arches would not collapse once the centering was removed. His great reputation and persuasive arguments won through, but there were still many skeptics who thought the bridge would collapse and kept up their protest many years after it was finished.

The five arches of Pont Neuilly are each 120 feet in span, and the piers are 13 feet thick, making a pier-width-to-span ratio of 1:9.3, a slenderness that surpassed all other arches by a huge margin. To build the pier foundation across the river, huge cofferdams were built, which were drained by using a bucket wheel

ABOVE: An artist impression of the Pont de la Concorde and the Place de la Concorde, commissioned for King Louis XV.

operated by a paddle, driven by the river current. The river bed in the cofferdam was then excavated to a depth of 8 feet below low water level and then piled down to the gravel layer using a drop hammer which was driven by two horses. A raft of piles were driven to refusal and then cut off 10 feet below low water level. An open grillage of timber beams was laid across the piles with the interstices filled with stones cemented together with lime mortar. Following on, the masonry for the piers was built up from this platform.

Next, the timber centering for the arches was built from the pier foundations for each of the spans. Once the big masonry abutments at each end of the bridge were finished, it was important to have all the arches finished in a single season between spring and fall, before the onset of the winter floods, which might wash away the centering. A total labor force of 872 men and 167 horses were engaged in transporting materials and building the arch voussoirs and spandrels for the bridge in the drive to get the arches completed. On September 22, 1772, in the presence of King Louis XV, all the centering was removed and the bridge was revealed. The Neuilly bridge was to stand in position for nearly 200 years, withstanding the floods and scour of the Seine, escaping damage from both World Wars. In 1956 it was demolished to make way for a larger bridge crossing. And so one of the most graceful and beautiful stone-arch bridges was lost to posterity in the name of progress!

Perronet was by now covered with honors and had became the first director of the Corps des Ponts et Chaussées, the first school of engineering in the world. The school and its first director, Perronet, secured the supremacy of French road- and bridge-building expertise throughout Europe. In the last 30 years of his life he was the chief of the Corps itself and publicly recognized as the premier engineer of the realm. But the culmination of his engineering achievement was not Pont Neuilly, nor his last bridge, the Pont de la Concorde: it was at Sainte-Maxence over the Oise, 35 miles north of Paris.

"Springing at a height of 18 feet from its tall and slender columnar piers (only 9 feet in diameter), it was truly a tour de force in stone arch construction," says Professor James Finch. It was the most slender and daring stone arch ever built, having a pier-width-to-span ratio of 1:12. The greater thrust from such a flat arch was carried by deep stone voussoirs, which occupied nearly all the space from the bottom of the arch to the roadway.

BELOW: "A great cheer went up when the centering was removed for Pont Neuilly Bridge in 1772."

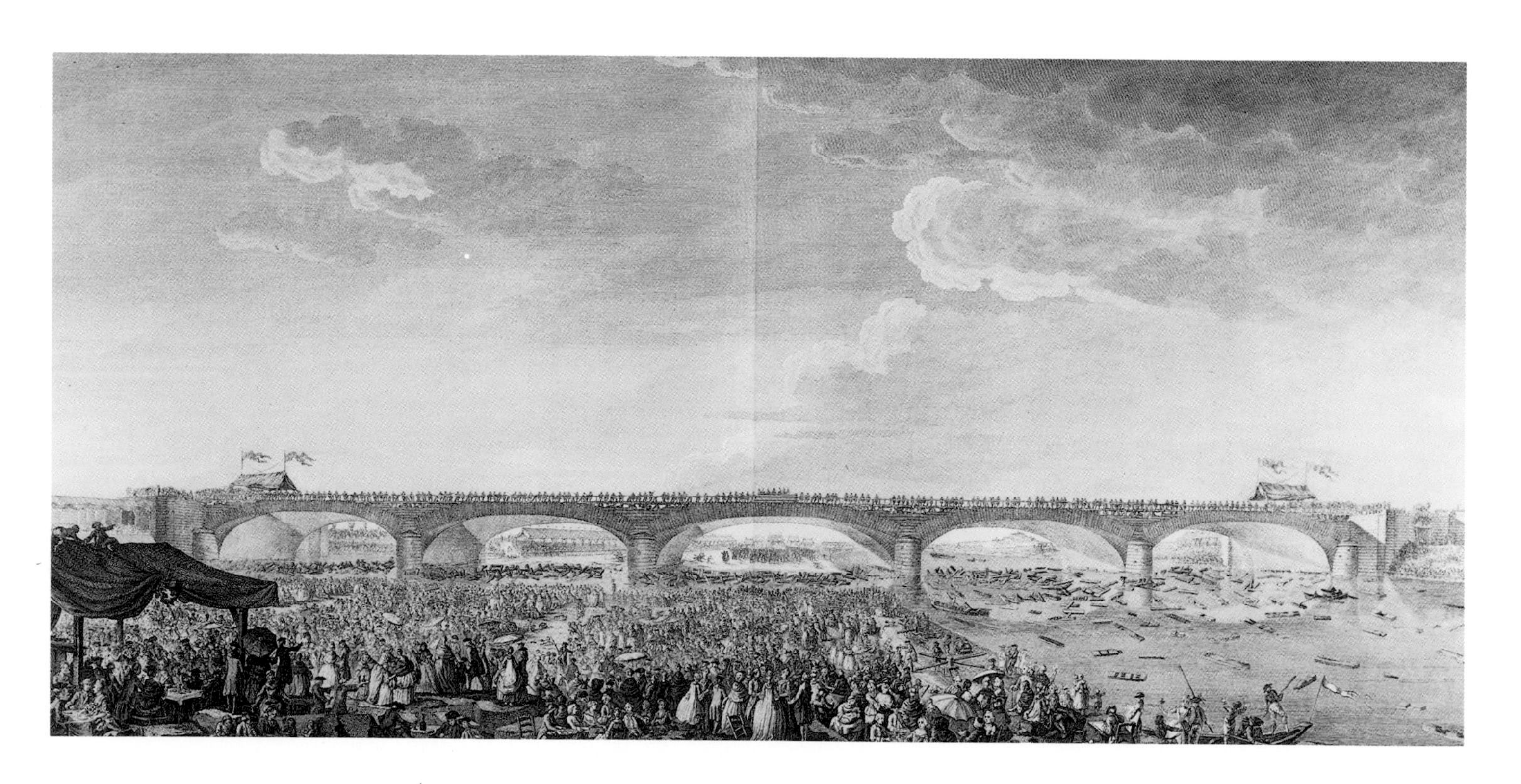

LEFT: The segmental arches of Pont de la Concorde as seen today.

And just as daring in concept were the piers for the bridge. They consisted of a pair of double columns connected by a lateral arch. The pair of outer columns were connected by bracing walls, while the inner pair were joined by the lateral arch. The bridge was partially destroyed in 1814 by Napoleon's retreating army, and was destroyed when the Germans retreated from the Marne in 1914.

To celebrate the work of Jean-Rodolphe Perronet, the "Father of Modern Bridge Building," why not take a visit to Paris and see his last and most famous bridge, the Pont de la Concorde, built in the shadows of the guillotine during the reign of terror of the French Revolution. It was on this bridge that Perronet died in 1794 at the age of 88 while supervising the construction, living in a small hut built for him at the end of the bridge by the militia. Could he have been oblivious to the gruesome activity going on within earshot of the bridge, as the crowds cheered and drums rolled, when another victim was felled by the guillotine? He must have known many people close to the Royal household, some of whom would have been sent to the scaffold.

The beautiful irony of the Pont de la Concorde is also its charm: it was commissioned by Louis XV and finished by the people of the French Republic; it was called the Pont de la Concorde by Louis XVI and renamed Pont de la Révolution by the Republicans; and while chaos and terror was spreading through France, on the banks of the Seine a team of workers united under him, went quietly about their daily business sifting the rubble of the Bastille to salvage masonry to place on the great bridge.

BELOW: Perronet's drawings for the building of Pont St. Maxonoc over the Oise.

DESSIN DU PONT PROJETTÉ SUR LA RIVIERE D'OISE ,
A PONT SAINTE MAXENCE

ABOVE: Thomas Telford.

BELOW: Buildwas Bridge, Coalbrookdale, England (1796).

Thomas Telford (1757–1834) *An engineer is a person who was innocent of specialization and expected to be able to build or repair any structure, material or man-made work. Telford was such a person.*

In 1777 Coalbrookdale, England, was the site of the first cast-iron bridge to be built. At the time this remarkable achievement did not attract great attention. Perhaps this was because it takes time for new ideas to become accepted, or perhaps it was because the bridge was an imitation of a masonry arch, with the cast-iron ribs taking the place of the stone voussoirs. But the second cast-iron bridge to be built some 15 years later and only three miles upstream from Coalbrookdale had quite a different reception. It was the first cast-iron bridge to express the true potential of the material.

Buildwas Bridge used only half the weight of material of Coalbrookdale and yet it was a flatter and more graceful arch. It was one of the many bridges that were to establish Thomas Telford as the greatest pioneer of the iron bridge structures in the world.

He was born in 1757 in the most humble of circumstances, into a poor family of hill farmers, living in a one-room cottage in the county of Dumfries in Scotland. He was brought up by his mother, because his father died when he was very

young. His mother relied on the goodwill of the neighbors to help with the small farm and to keep them from starvation.

More or less self-educated, with a cheerful disposition and a warm smile, the boy who kept sheep for relatives and ran errands for his neighbors was to become the greatest civil engineer in the country and the builder of the first iron-chain suspension bridge. After attending the village school he was apprenticed to a stonemason at age 14 and assigned to work on the estate of the Duke of Buccleugh in Langholm. There was plenty of work for young Telford, in building cottages, public buildings, dams, and bridges for the Duke, who was the richest landowner in all Scotland.

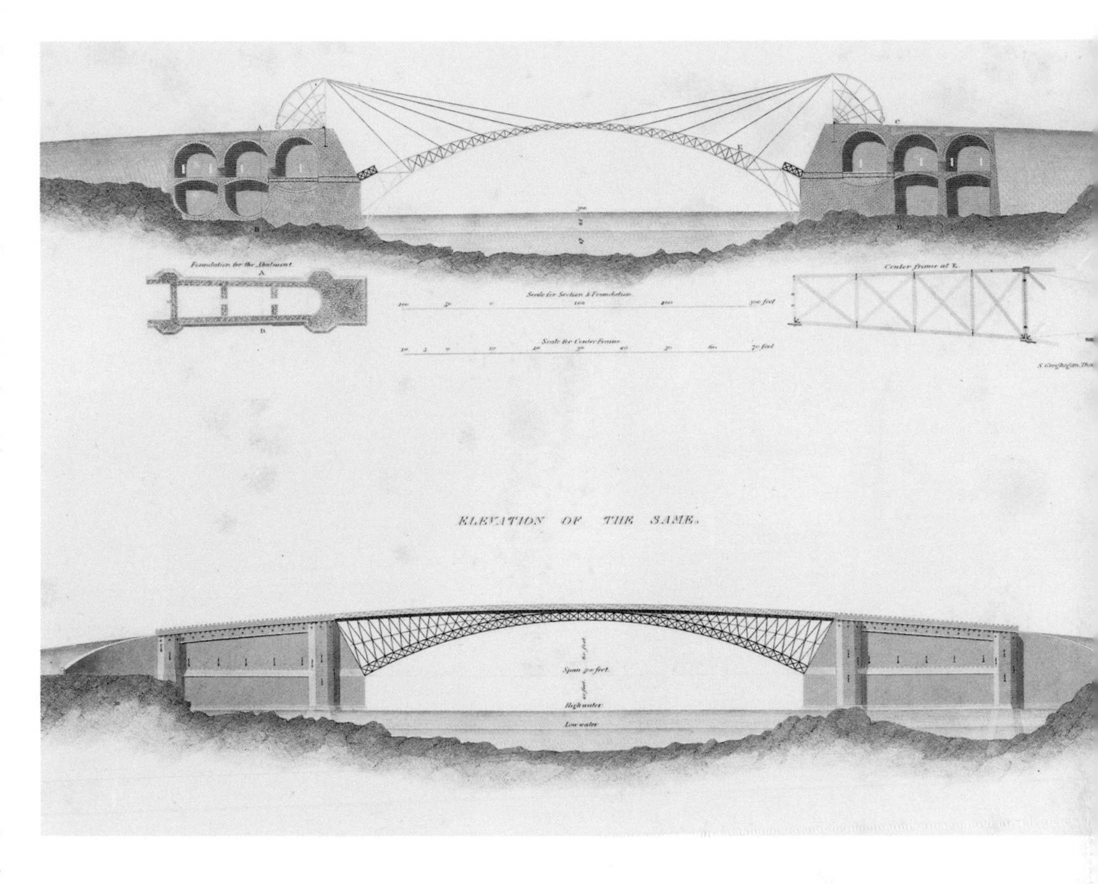

ABOVE: Telford's engineering drawing for a later iron truss arch.

After serving his apprenticeship he left Langholm for Edinburgh, where the pay was better and there was plenty of work to do. Two years later at the age of 25, full of confidence in his ability as a stonecutter and master mason, armed with letters of introduction from his Scottish patrons, he set off to find work in London.

Telford eventually secured work with the architect Sir William Chambers on the construction of London's famous Somerset House, starting as a lowly hewer but soon working his way up to become head mason. Telford's letters home to his mother give an insight into the mentality of his fellow workers and Telford's ambitions. According to Telford there were two distinct types of workers on the site—the majority of them just worked for their week's pay and a chance for a bit of fun after work, while Telford on the other hand was conscious of building a career and studying in his spare time. Telford from all accounts was a sociable, gregarious person with a lot of friends, but he never married.

In 1784 Telford left London to work on a project for the navy in Portsmouth Docks. Within a year he was promoted to general superintendent, and he then embarked on a long and challenging career as a civil engineer, first working as an engineer for the Ellesmere Canal Company, where he designed a series of aqueducts that included the first all-iron aqueduct at Longdon-on-Tern on the Shrewsbury Canal in addition to the famous Pontcysyllte Aqueduct. The Pontcysyllte remains unique in scale and magnificence, reminiscent of the grandeur of the Pont du Gard, and carries the Ellesmere Canal over the Dee valley near Llangollen in Wales.

Nineteen cast-iron arches, each spanning 45 feet, 127 feet above the river bed, are carried on stone piers, to support the cast-iron aqueduct for 1,000 feet over the gorge. It was during this period that Telford settled in Shrewsbury, in the heart of the iron and coal industry, and watched with interest as Abraham Darby and John Wilkinson tried to work out the problems of building a cast-iron arch. He studied the plans for another tremendous cast-iron bridge that was being built by Tom Paine in America with a span of 400 feet over the Schuylkill river, not far from Timothy Palmer's covered bridge. Paine had the giant cast-iron ribs forged in Coalbrookdale but, shortly after they were formed, his financial backers pulled out and the project was scrapped. The iron was sold as scrap and was used for building an iron-arch bridge over the Wear at Sunderland in 1796.

RIGHT: Beautiful Craigellachie Bridge over the Spey in Scotland (1815).

Now a resident in Shropshire, Telford took up the position of County Surveyor of Shropshire on the personal recommendation of Sir William Pulteney, who had employed Telford while he was in Edinburgh and when he first went to London. As County Surveyor, Telford was required to design many roads and canals, and to work on bridge construction and even railroads. Where did all this new knowledge and expertise suddenly come from? Telford was always studying books and buildings whenever he could find the time—visiting old cathedrals, reading Vitruvius's books on architecture, studying Wren's designs and Perronet's books on bridges no doubt—on every related subject in the field of civil engineering. He had already helped to build masonry bridges while serving his apprenticeship in Scotland and had a good grounding in cast-iron construction, being so close to Coalbrookdale, and so it was natural for him to try his hand at a cast-iron bridge when the old bridge at Buildwas was washed away by a flood

ABOVE: Craigellachie Bridge, the first iron truss arch bridge.

in 1795. Buildwas Bridge was the first cast-iron bridge to show the world how the material could be used to greater advantage than masonry-arch construction. It was lightweight, spanned farther, and could be built out from each abutment without obstructing the waterway! Craigellachie Bridge, built in 1815 over the Spey, was Telford's masterpiece and the most beautiful iron bridge built by him; it was also the first metal-arch-truss structure that did not try to imitate the masonry arch and spandrel. Many famous bridges were built in England during the next two decades following Telford's mastery of cast-iron construction.

The career-minded Telford went north to design and supervise the building of the Caledonian Canal in Scotland to eliminate the laborious and perilous sea journey around the north cape of Scotland. Then in 1810 he was engaged in the Holyhead Road Survey, which aimed at improving communication between London and Ireland by developing a good road link between Anglesey (off North

Wales) and mainland Wales, for the new harbor at Holyhead. The great problem was bridging the Menai Straits, between Anglesey and the mainland.

John Rennie, one of Telford's great contemporaries, had suggested an iron bridge with a single span of 450 feet, but the cost of construction was prohibitive at £290,000 (approximately $470,000) in those days. Meanwhile Telford was being consulted on his ideas for a bridge crossing of the Mersey at Runcorn, and, having read about designs for suspension bridges in America and France, astonished his clients when he said to them, "I recommend a bridge of wrought iron, upon the suspension principles."

Telford knew that suspension bridges were notoriously flexible and unstable, although recent designs by James Finlay in America had been built with a level bridge deck and were meant to carry road traffic. But, even then, one of Finlay's bridges had collapsed under a drove of cattle and another under snow and ice. The suspension principle was simple in theory, it was cheap on materials, but it was more treacherous in practice than engineers in the 1820s realized. But Telford had the instinct to know that he could solve this flexibility problem.

Telford collaborated with a Captain Samuel Brown, who had become interested in suspension bridges, and had invented a new flat-iron chain which he believed capable of supporting heavy loads without deforming. They became good friends and worked together on a series of tests for a chain-link suspension bridge. The Mersey crossing did not go ahead because the cost was too expensive, but all was not lost. Much of the test work and research that Telford had carried out with Brown was to prove invaluable for Telford's biggest challenge.

The Menai Straits Bridge is Telford's great memorial, and secures his place in engineering history. And, though he was nearly 70 when it was completed, he continued on in his career almost to the day he died at the age of 77 in 1834. The

ABOVE: The 1000 ft long cast iron aqueduct of Pont Cysyllte is supported on brick arches.

BELOW: The Chirk Aqueduct built in 1801 carries the Ellesmere Canal over the Ceriorg Valley in Wales.

OPPOSITE: Pont Cysyllte aqueduct in Wales (1805), reminiscent of the grandeur of the Pont du Gard.

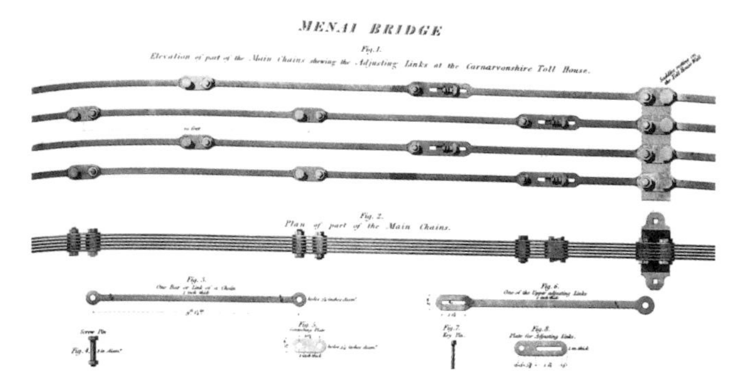

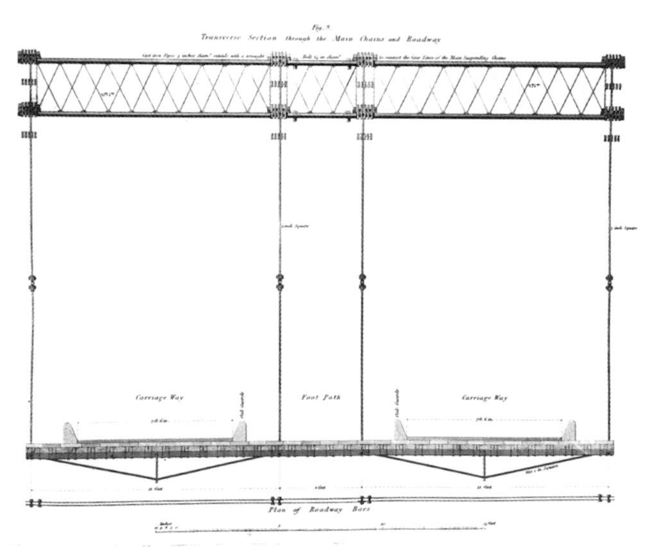

ABOVE: Detail of the chain link bridge deck structure and hangers, taken from drawings prepared by Thomas Telford's office.

bridge required 2,000 tons of wrought iron to build, which was a considerable amount, but it was far less than the 6,000 tons needed for the iron-arch spans of Vauxhall Bridge over the Thames in London. Through 1820 to 1822 work on two-pier foundations and towers continued, alongside testing of the flat-iron bars, 935 of which were needed to make each of the bridge's 16 cable chains. The two great suspension towers, which were hollow over the top third, were raised 30 feet above the roadway level and completed in the spring of 1824. The carriageway, which narrows to 9 feet to pass through the towers, was built in sections on the ground and fitted with cast-iron plates and saddles for connecting them to the chain cable. The rest of the year was spent getting preparations ready for suspending the first cable chain.

First a half-section of the chain on the mainland side was pulled over the tower and draped down the tower face, until it met a barge in the water. Then a portion of the chain was laid over the top of the Anglesey tower. Next a barge was floated out with the final section of the chain cable and one end was then attached to the existing chain while the other was fitted to heavy ropes pulled by a capstan on the other bank.

Telford's diary records the final moments: "... the said ropes passed, by means of blocks, over the top of the pyramid [tower] of the Anglesey pier. Then the workmen who manned the capstans moved at a steady trot, and in one hour and thirty-five minutes after they commenced hoisting, the chain was raised to its proper curvature, and fastened to the portion of chain previously placed at the top of the Anglesey pyramid."

Within 16 weeks the last of the 16 chains were lifted into place. The bridge deck, the toll gates, and side railings were all in place, and so, finally, was the road, with three thicknesses of fir planking, each layer separated by felt.

On January 30, 1826, the London mail coach occupied by Telford's assistant engineers, the mail-coach superintendent, and anyone else who could find a foothold on the coach crossed the world's first bridge over an ocean and the longest span in the world at 580 feet.

The astonishing feature of many of Telford's bridge structures was the relative ease of their construction, the lack of accidents, and the certainty that they could and would be built. Telford, by all accounts, was a good-humored man, an entertaining raconteur, an intuitive and brilliant engineer, and a lively writer, who kept a detailed diary of events and wrote regularly to his friends in Scotland and to his mother unfailingly until her death. He never married but in his letters there was often reference to a woman, an actress whose talent he admired. The explanation given by his close friend and biographer Rickman was simply "he lived as a soldier; always in active service!"

ABOVE: The Menai Straits Bridge as it appears today.

OPPOSITE ABOVE: A painting of the completed Menai Straits suspension bridge not long after it opened in 1826.

LEFT: "The first bridge to be built over an ocean" was also the longest span in the world at 580 ft when it was completed.

ABOVE: John Roebling.

The Roeblings: John A., Washington A., and Emily Warren—a magnificent obsession

> But a bridge is more than the embodiment of the scientific knowledge of the physical laws. It is equally a monument to the moral qualities of the human soul ... Let us then record the names of the engineers who have thus made humanity itself their debtor ... They are John A. Roebling, who conceived the project and formulated the plan of the bridge; his son Washington A. Roebling ... who directed the work from its inception to its completion upon the tragic death of his father ... and one other name which may not find a place in the official records, but cannot be passed over in silence ... this bridge is an everlasting monument to the self-sacrificing devotion of Mrs Emily Warren Roebling ...

So said the Honorable Abram S. Hewitt, civic leader of New York, during the inaugural speeches that marked the opening of the Brooklyn Bridge in 1874.

During the ceremony, a lonely paralyzed man, crippled and racked with pain, sat at an upper window of an apartment block in Columbia Heights, viewing the scene on the bridge through field glasses. Through them he saw the distinguished procession coming over the bridge, which included the President of the United States, the governor of the state, and many other notables, together with a glittering military escort.

His throat was choked with emotion, and he could barely hold back the tears, for this was the greatest moment of Washington Roebling's life. This day gave

RIGHT: Brooklyn Bridge. This picture was taken in 1916.

ABOVE: Emily Warren Roebling, the wife of Washington Roebling.

meaning to his suffering and his father's life and dedication before him. This was his father John Roebling's bridge, which he had built against all the odds.

Beside him at the window stood his wife, Emily, who had been his ministering angle through the years of pain and struggle that he had endured. She had been his eyes and his feet for ten harrowing years, recording his notes and carrying his every instruction to the workmen in the caisson, as he lay paralyzed from caisson disease (the bends). He could look back now with immense pride at the Brooklyn Bridge, at what he, his wife Emily, and his great father John Roebling had striven to achieve.

Suspension ideas John Roebling came to America in 1831 at the age of 25, not in search of God or bread, but to make a living as a civil engineer and bridge builder. Roebling was the son of a middle-class German family, with an excellent education at the Royal Polytechnic Institute of Berlin. He was traveling on a ship heading for Philadelphia with $400 in his pocket, which was the equivalent of a small fortune in those days.

Once landed, Roebling, his brother Carl, and his compatriots headed for the frontier lands of western Pennsylvania, to set up farmlands and to establish a community, which they called Saxonberg. Roebling kept up his engineering studies in the evenings while doing farm work in the daytime, and waited for an opportunity to get started as a civil engineer.

In 1837, through a good friend, he got a job with the Sandy and Beaver Canal Company. Two years later he was surveying the route of the railroad from

BELOW: Niagara Bridge built in 1851.

ABOVE: Washington Roebling.

Harrisberg to Pittsburgh and that same year he had his first practical ideas on bridge building.

A feature of the canal system was the mechanism used to tow flatboats on rails, to the top of a steep hill or incline, and then put them back into the canal on the other side. Heavy hemp cables were used to haul the boats, but they frequently broke and on occasion caused fatal accidents. Roebling thought about using wire rope, an invention he had just read about in a German technical magazine. No one in America had ever *seen* a wire rope, let alone heard of its being used, and naturally the canal company did not think his idea would work.

Roebling decided there and then to set up his own wire-making company with machinery he could develop, but he needed people to make it. He called on his Saxonberg neighbors and willingly they turned themselves into an efficient team of workers.

Roebling's wire rope was stronger and easier to handle and outlasted the best hemp rope by an irresistible margin. With his civil-engineering skills he soon established himself as one of the leading canal engineers of the region, designing, building, and repairing aqueducts over rivers using suspension structures supported by wrought-iron cables spun by his company. The public and the technical press were much impressed with the daring suspension structures designed by the inventive Roebling, but he wanted to go on and build bigger structures, such as road and rail bridges.

He got his big break in 1851, when the mercurial Charles Ellet quit as engineer for the Niagara Falls Bridge, and Roebling was invited to take over.

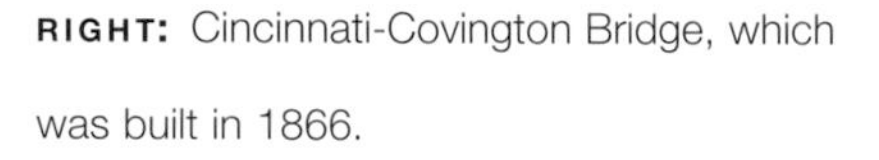

RIGHT: Cincinnati-Covington Bridge, which was built in 1866.

History is made at Niagara Roebling's Niagara Bridge is a genuine landmark in bridge-building history and established the pedigree of the suspension bridge while making a fortune for Roebling and his wire-rope company. It was a robust structure that was light and strong, and carried a railroad on the upper deck and a roadway on the lower. It used just a quarter of the total iron tonnage that was used to build Robert Stephenson's great Britannia Bridge over the Menai Straits in Wales.

The Niagara Bridge actually supported a bigger load than the Britannia and spanned 861 feet to the Britannia's 460 feet. It was a fantastic achievement. After it was built Roebling was moved to say that "bridges of half a mile in span, for common or railway travel, may be built using wrought iron for cables with entire safety." Some visionary! Some engineer! And how right he was to be!

Other bridge commissions soon followed, and, by the time he embarked on the design of the record-breaking span of the Cincinnati and Covington Bridge, he had the assistance of a bright young engineer, his son Washington. He was one of the first graduates of Rensselear Polytechnic Institute, one of the pioneer engineering colleges in America. Although father and son were quite different in character and temperament—Washington was engagingly open and American while John was reserved and stoically German—they both shared a passion for building suspension bridges.

The American Civil War interrupted the father–son partnership for a while. Colonel Washington returned to his father's bridge-building company a highly decorated officer, not only for gallantry in battle but for building military suspension bridges in record time such as the ones at Harper's Ferry and Fredrickberg. In a way this toughened Washington's mental fiber for the great ordeal that destiny had in store.

The Brooklyn Bridge In May, 1876, less than six months after Washington had returned to help his father complete the Cincinnati Bridge, John Roebling was hired as chief engineer for the design and construction of the record-breaking span of the Brooklyn Bridge. For two years father and son worked together to complete the detailed drawings and plans for the bridge. Then, tragically, one day while John was working on a survey pontoon to determine the location of the Brooklyn tower, his foot was crushed when a ferryboat struck the pier he was standing on. Despite

RIGHT: The East Tower seen in 1998.

FOLLOWING FOUR PAGE SPREAD: Dawn light and mist frame the Brooklyn Bridge; the bustling aerial view by day.

BELOW: Erecting the truss deck from the suspension cable on the Brooklyn Bridge.

PIER 17

the amputation of his foot, John Roebling died of tetanus less than three weeks later, at the age of 63, in 1869.

With John Roebling gone a new guiding spirit was needed to carry the great work forward. On his deathbed the great man turned to his son and asked him to complete the task of building the bridge. Courageously and loyally, Washington Roebling now assumed the gigantic responsibility of keeping the construction moving forward and of finishing his father's greatest dream.

During the days and nights that the work was going on under the bed of the East River, hand digging through the slush and mud in the dank, dimly lit compression chamber of the caisson, Washington was in and out of it many times a day. He directed the efforts of the men, setting an example of courage and commitment. Night and day he worked with the men under the crushing, ear-splitting air pressure of the caisson and the foul smells of putrid soil and waste.

LEFT: Snow covered pedestrian walkway of the Brooklyn Bridge.

Then, one afternoon in the early summer of 1872 on climbing out of the caisson, he collapsed and had to be carried out. He was a victim of caisson disease, with his death expected that evening. He rallied but remained partially paralyzed and in great pain. He was doomed to a life in a wheelchair and his bed, suffering in pain as progressive blindness and deafness began to set in.

Thankfully, his brain and his hands were active, as was his will to live to see the job done. Every difficult or specialized erection procedure of the bridge he anticipated and calculated with meticulous precision, issuing drawings and sketches with specific instructions. In this manner the work continued without a serious hitch for the next ten years. Although there were many occasions when the board wanted to retire him from running the project, he was always able to convince them that, with his wife Emily at his side, he could continue to do so. Under his guidance Emily studied higher mathematics, the calculation of catenary curves, the strength of materials, stress analysis, and bridge engineering. She was quick to learn, picked up her husband's ideas fast, and was soon a competent civil engineer.

She made daily visits to the bridge, to the caisson, the tower, and cable yard, to inspect the work for her husband and to carry out his instructions to his staff of engineers. She was his coworker and principal assistant, his messenger, and sole contact with the outside world.

The Brooklyn Bridge will always be associated with one name in the history of bridge building. In Emily Roebling we remember all that is admirable in the human spirit and all that is noble in the construction of a bridge. Washington Roebling unexpectedly lived on till he was 89, outliving Emily, and even partially recovering from his paralysis.

ABOVE: Cable spinning on the Brooklyn Bridge—steel sections of the truss deck can be seen stacked in the foreground.

LEFT: The East Tower seen in 1998.

Othmar H. Amman (1881–1967) Amman was born in Switzerland in 1881, in the village of Schaffhausen, where many years ago Hans Grubenmann built a famous truss bridge. Perhaps the timber bridge inspired the young Amman, who wanted to be an architect but was encouraged to study engineering at Swiss Federal Polytechnic Institute in Zürich, because he was gifted at mathematics.

He started work as a draftsman in an engineering firm in Germany, but soon became bored and frustrated, because his ambition was to be involved in building the biggest bridges in the world. The place to realize his ambitions was America, where the great Brooklyn, Quebec, and St Louis bridges had been built and where many more long-span bridges were being proposed.

ABOVE: Othmar Amman aged 70.

He sailed for New York in 1904 and secured a job with a consulting engineer, Joseph Meyer. He then went to work with Frederick Kunz, whose Pennsylvania Steel Company was involved with the Queensboro Bridge in New York, a long-span cantilever truss. Then the next change came shortly after the Quebec Bridge collapse in 1907, when Amman offered his services, for no pay, to work with C.C. Schnieder, whose company was appointed to investigate the failure.

And it was in this roundabout way that the restless Amman met up with the famous Gustav Lindenthal, whose company were also advisers to the Canadian government on the Quebec Bridge collapse. At last he had found a practice that was to stretch his design talents, that suited his ambitions, and in Lindenthal he had a mentor he really admired. He worked for Lindenthal in all for ten years, with a break of a couple years when he was recalled to do army service in Switzerland during World War I.

When he first joined Lindenthal his first assignment was as first assistant in charge of office and field operations of the Hell Gate Bridge, which was about to begin construction. Unfortunately, Amman had to leave for army service in 1914, and did not see the Hell Gate finished. Upon Amman's return to America,

BELOW: The Bayonne Bridge over Kill van Kull in New York, arguably the finest truss arch span ever built (1931).

Lindenthal placed him in charge of designing a railroad bridge across the Hudson at 57th Street in New York. In the end the project was not built because it did not attract enough sponsors, but Amman did not give up on the idea of building this much-needed bridge and thought that it might be lighter and cheaper to build if it was erected at some other point. According to Margaret Durer Amman, Othar Amman's daughter, now aged 76, her father pointed out that a railroad bridge over the Hudson at its widest point was too expensive to finance. "He had suggested to Lindenthal that it would be better to build a bridge further upstream, roughly where the George Washington is today, and to make it suitable for motor cars and not a railway, as this was going to be the transport of the future. They argued over this for two years. No one was willing to put money into the scheme. In the end my father grew extremely frustrated and resigned. Governor Silver of New Jersey got to hear about my father's concerns and approached him about becoming Chief Engineer of the newly formed New York Port Authority," says Margaret Amman. "Of course Governor Silver and my father go back a long way, to a time when Lindenthal was nearly bankrupt, my father was penniless and I had just been born. You see after the Hellgate bridge Lindenthal had no more work, and slowly as the funds ran down he had to lay off every one of his staff including David Steinman, except my father, whom Lindenthal thought the world of." Lindenthal offered Amman a post of managing a clay mine in New Jersey, which he and Silver had bought at a knock down price, some years ago. It had been losing money. "The job came with a house and a big garden, so my father took it … there was no other work." Within two years Amman made that clay mine into a profitable pottery business and earned Lindenthal and Silver a small fortune for their minimal investment. Silver had not forgotten that.

TOP: The truss of the Bayonne Bridge, New York.

ABOVE: Father and daughter (Margaret Amman-Durer) together in 1924.

Meanwhile, Amman continued to design other successful bridges for Lindenthal until 1925, when he left to become bridge engineer for the newly formed Port of New York Authority. The new public corporation was set up with the purpose of purchasing, constructing, and operating terminals and transportation facilities—which, broadly translated, meant lots of bridges.

In the course of the next six years, Amman, with the help of his assistant Alton Danna, was to design two magnificent, innovative, and record-breaking bridge spans, whose names are as famous as the cloud-busting skyscrapers on the Manhattan skyline. The record-breaking span of the mighty George Washington Bridge—a leap of 89 percent more than the previous record of the Quebec Bridge—was opened three weeks before the great steel-arch truss of the Bayonne Bridge over the Kill Van Kull in November 1931. The former was the longest span while the latter was the longest steel-arch span in the world. It was unprecedented that two structures of such scale had been undertaken and completed simultaneously by one organization and one man.

The Kill Van Kull waterway, which separates New Jersey from Staten Island in New York, is an estuary about a quarter of a mile wide. Through this waterway passes most of the cargo arriving and leaving New York, which in tonnage handled was greater than the Suez Canal at its peak. Any bridge that was proposed would need a clear span of 1,650 feet with no intermediate supports to obstruct or narrow the navigation channel.

ABOVE: Amman shaking the hand of Howard Hughes watched by Nelson Rockerfeller at the dedication ceremony when the second bridge deck was added to the George Washington Bridge.

LEFT: The tower of the George Washington Bridge (1931).

ABOVE: Goethals Bridge, New York (1929).

Amman's experience and knowledge of cantilever and steel-truss design was second to none and he also knew a lot about suspension bridges. From topographical survey maps of the areas, followed by borings and soundings, a thorough study of the ground conditions was prepared. This revealed that the estuary was underlain by good rock formation not too deep below low water level.

The choice for Amman was interesting, because he could design a cantilever-truss, suspension, or steel-arch bridge without difficulty for the clear span. In the end it was economic and aesthetic considerations that dictated the way forward. The first option, the cantilever-truss bridge, would be bulky and require more steel to construct than the other two, while the second option, the suspension bridge, would need deep and expensive excavations in the bedrock for forming the anchorages for the cables. The dense rock that made the suspension bridge expensive was in fact ideal to carry the large thrusts from an arch span.

Amman designed the Bayonne as a two-hinged, parabolic, through-arch truss, whose main ribs are the bottom chord of the arch and rise 266 feet from the water line. The top chord of the arch acts as a stiffener to the bottom chord. The central portion of the bridge deck below the arch is suspended from the trussed arch by wire rope hangers. The steelwork was supplied and erected by the American Bridge Company, who offered to use a new and cheaper carbon-manganese steel rather than nickel steel for the main chord. Amman ordered a series of tests on the new steel to check its suitability as a replacement material, and, although it was not as good in tension as the nickel steel, after some design modification it was accepted. Over 5,000 tons of it was used on the Bayonne Bridge, the first bridge to use carbon-manganese steel, which is the composition of modern steel.

The abutments that transmit the arch thrust directly to the bedrock are close to the water line, while the lightweight steel towers at the end of the arch span have been left unclad, just like the towers on the George Washington Bridge. Cass Gilbert, the architectural adviser to the Port Authority, detailed masonry to clad the towers, but it was omitted on economic grounds.

Controversy still surrounds the towers of the Bayonne. Would they have looked better clad in masonry, like the monumental towers of the Hell Gate and Sydney Harbour Bridges? Undoubtedly the bridge as it was built expresses the purity of its engineering design, and its lightweight construction. Cosmetic camouflage on this occasion would surely have been an aesthetic blunder.

Amman went on to design a number of other notable New York landmark bridges—the Triborough, the Throgs Neck, the Goethals, and the Outerbridge. But it was Amman's Bronx Whitestone suspension bridge, completed in 1939, that was universally acclaimed as the ultimate in suspension-bridge design. The slender plate-girder stiffeners for the deck span, and simple yet elegant towers, represented the very antithesis of nineteenth-century bridge construction. Just

BELOW: Verazzano Narrows Bridge, New York built in 1965.

OPPOSITE: Looking down the face of the tower of the Verazzano Narrows Bridge.

when Amman and his great rival David Steinman—who between them had designed most of the suspension bridges of North America—felt they had mastered the science of the long-span suspension bridge, Mother Nature dealt the bridge world a cruel lesson.

A year after the Bronx Whitestone Bridge opened, the Tacoma Narrows Bridge collapsed, causing bridge engineers the world over to re-evaluate their approach to suspension-bridge deck design. The Bronx Whitestone deck was rebuilt with a conventional heavy truss superimposed over the slim plate girders. Diagonal stays were also added, which did nothing to help the bridge aesthetics, while the Golden Gate underwent a $3,500,000 strengthening. Overnight, the heavy truss-girder bridge deck, which had been replaced by the slim plate girder, was brought back into suspension-bridge construction.

Amman's great rival Steinman beat him to win the commission for designing the big Mackinac suspension bridge in 1950, but Amman was to have the last laugh. In 1959 the Triborough Bridge and Tunnel Authority appointed Amman to design the colossal Verazzano Narrows Bridge, the longest span in the world. It was a scheme first proposed by Steinman in the 1920s, but after years of political debate he was never given the go-ahead.

It began two years after the Mackinac was completed and was the mightiest suspension bridge of all. Its size was matched by its huge expenditure: $325 million. No bridge had cost so much or used so much material before. The bridge was financed out of public funds and the cost recovered through car tolls, which has proved to be an extremely sound investment, since the traffic volume over the bridge doubled within eight years, going from 10 million to a staggering 25 million vehicles.

The tower rises 650 feet above the water line, the deck is suspended 226 feet above it, and the main span stretches a phenomenal 4,260 feet between towers, dominating the skyline and the waterway from Brooklyn to Staten Island. "Everything about the Verazzano Narrows Bridge," observed the *Engineering News Record* "is big, bigger, or biggest!" The four steel cables alone cost more than the whole of the Golden Gate Bridge, and there is enough strand in the cable to wrap themselves around the world a mind-boggling six times. The cables must support a dead weight of 120,000 tons from the bridge, plus the cable's own weight of 39,000 tons.

Despite the view that the Verazzano was not an outstanding bridge design, it is visually so impressive that even the heavy double-deck roadway looks in proportion when juxtaposed between the mass of the towers and the open span. It is a very photogenic subject, as demonstrated to great effect opposite, and is arguably one of the most impressive bridges that straddles New York's waterways. The year after the Verazzano Narrows Bridge was opened on September 22, 1965, Othmar Amman died at the age of 86.

Eugene Freyssinet (1879–1962)

> There is not any field of constructive activity—and I say not any, after close consideration—to which the idea of prestressing does not provide, often, unexpected possibilities.
>
> *Eugene Freyssinet*

I can best introduce Eugene Freyssinet to you by beginning this profile with an extract from an article called "His Own Man," written by Sir Alan Harris in *Concrete Quarterly* in 1992:

> What was he like? Small in stature, with a pussy cat face, he was capable of tigerlike roars, but his own angers were soon forgotten: all he asked for was complete devotion to the job! He had no taste for social life. At a celebration at the Ecole des Ponts et Chaussées to honor his 60 years, I found myself sitting next to him in the front row of a lecture hall faced by a stage full of the great and the good in French civil engineering, all waiting to speak in his praise. "Watch that lot," he whispered to me, "every time prestressing is mentioned, each looks as if he had a kick up the backside!" There was once a garden party at his house and it was a complete disaster, but he did turn up at my wedding in Paris and was charm himself.

ABOVE: Eugene Freysinnet in his 50s.

The early years Freyssinet started as a junior in a local office of the Ecole des Ponts et Chaussées at Moulins, where his job was adviser to a number of rural mayors. He loved his work, knew his clients and their needs, and was given total freedom to design structures of his own devising, usually in reinforced concrete, which the mayors organized to have built by local labor. "Anyone," he said, "who told them that those bridges were contrary to the regulations would have run a heavy risk."

Many of those bridges are still standing today. The three identical bridges over the Allier just below Moulins are the most impressive. They were of three spans of 238 feet each, with very flat concrete arches. He built a test arch bridge using prestressed concrete as a foundation tie and discovered the existence of creep in

RIGHT: Boution Bridge over the Alliers (c1912).

concrete. This discovery was to spark his early enthusiasm and later dedication to the art of prestressed concrete.

He first thought of it in 1904 but it took him another 20 years to bring prestressed concrete to full practicality. In ordinary reinforced concrete the liquid concrete is poured into forms or molds around the reinforcing bars. As the concrete hardens the formwork bears the load. Not until the forms are removed does the concrete come under any stress. This is the stage when the loading, combined with shrinkage as the concrete dries and temperature contraction as the concrete cools, does most damage and when cracks begin to appear in the tension zone, because concrete is weak in tension. Freyssinet's idea was to induce into the concrete, before erection or before the forms were removed, a precompression stress that would effectively counteract those damaging tension stresses in the finished structure.

BELOW: Airship hangars designed by Freyssinet at Orly Airport, 1921.

Freyssinet received a special prize and much publicity when the second of the three bridges, the Pont Le Veurdre, was built in 1911. Suddenly he was a celebrity in the bridge world at the age of 30 and was receiving approaches from clients all over the world to design and build bridges for them. He resigned from state service and set up as a designer-contractor with two partners.

Then came World War I, when Freyssinet was seconded to the Army Engineer Works, where he explored and exploited the use of concrete, building shell-roof industrial buildings, many bridges, and a number of seagoing cargo ships made of concrete. By the end of World War I he was a master of concrete construction and a pioneer of concrete shell-roof construction and returned to his business to design and build all manner of concrete structures—from airship hangars and industrial complexes to record-breaking spans for concrete bridges that were cheaper than any other in the market—culminating in the innovative Plougastel Bridge over the Elorn. Plougastel's record-breaking concrete arches with three spans of 590 feet, which were the longest in the world at the time, and the unique method of fabrication and flotation of centering established Freyssinet as one of the greats among bridge engineers. It was while building Plougastel that Freyssinet found the answer to creep and how to design and build structures with prestressed concrete.

ABOVE: St Pierre du Vouvray, 1922.

ABOVE RIGHT: Centering for concrete arch of St Pierre du Vouvray.

BELOW: Prestressed beams of the Luzancy Bridge (1945).

A life of prestressed concrete Freyssinet was determined now to make prestressed concrete his vocation, and left his successful practice, much to the disappointment of his two partners, who thought he was mad to abandon everything for such a risky new venture. He was 50 and was prepared to devote all his private fortune to developing prestressed concrete.

After a period of waiting, a client emerged who wanted to supply 40-foot concrete pylons for power lines that would be needed for a major expansion of the electric power grid. The demand for power lines would be so large that Freyssinet and his backer would need to invest in five or six factories located around the country. Shortly after the pilot plant was set up and the first batch of the precast concrete hollow tubes with walls just under half an inch (or 10 mm) thick were produced, the market collapsed as the depression of 1933 swept through France and Europe.

His factory was sold for scrap and he was ruined. He and his wife were virtually penniless. And then he made the bravest gamble of his life. The Gare Maritime, the largest shipyard building in France, located at Le Havre, was suffering with disastrous settlement and sinking. The old quay walls were fine but not the building. The shipyard was the intended dry dock for building the SS *Normandie*, which was the finest and most beautiful ocean liner to have ever crossed the Atlantic.

To offer to remedy the problem was bold, and at the same time to fail would be financial disaster for an eternity. But what had Freyssinet to lose? His solution of jacking up the Gare Maritime building was accepted, though he had never used this technology before.

ABOVE: Pont Annet, one of a family of five prestressed concrete beams bridges designed by Freysinnet over the river Marne (1947–50).

First he formed a prestressed concrete beam across the pile heads of the existing foundations. Then through holes made in the beams he used hydraulic jacks to push tubular-concrete piles, cast in lengths of about 6 feet 6 inches, down into the gravel layer, which commenced at some 33 to 50 feet below ground. When the concrete piles had reached a dense gravel layer, the subsidence slowed down.

He then used the piling jacks to restore the building to its intended level. Freyssinet was hailed as a great engineer and was congratulated by the leading civil engineers in France, including Edme Campenon of Campenon Bernard, who immediately engaged Freyssinet as their designer and championed his ideas for the rest of his life.

Prestressing was now established and, through the international reputation of Campenon Bernard, the stage was set for Freyssinet. Dams in North Africa were built, pressure pipes in concrete manufactured, many shell-roof structures and bridges were erected, precast bridge beams for mass production were granted licenses for manufacture in Germany and Switzerland.

Among the most important projects that were built was the family of five prestressed bridges—the Esbly, Annet, Trilbardou, Ussy, and Chagis—erected over the Marne between 1947 and 1950. According to Edme Campenon, this was the clearest souvenir of Freyssinet's creative genius.

In the first place, the choice of the structure—a raked leg portal bridge—connected with the obligation of leaving a wide navigable channel. Then the design had to fit between the profile of the carriageway and the soffit, which had a limited headroom. The construction method of placing the precast deck elements with an overhead runway, combined economy, speed of construction with a total absence of obstacles to navigation during erection. And finally the structure exerted thrusts on the abutments of the previous bridge which we were utilizing but which were far above those for which they were originally designed. Freyssinet found a way to re-use the old abutments by strengthening the soil behind the abutment by preloading the soil—it was soil prestressing.

RIGHT: Orly Airport Viaduct, one of the last bridges designed by Freysinnet.

Fritz Leonhardt (b. 1909) *In the 1950s while bridge engineers in America and Britain were preoccupied overcoming the problems of suspension bridges in high winds, in Germany a team of bridge engineers were developing a new type of suspension bridge—the cable-stay bridge. The leader of this group and the most honored bridge engineer living today, was Fritz Leonhardt.*

Leonhardt was born in Stuttgart in 1909 and enjoyed the discipline of school life at the excellent Dillman Gymnasium, from where he went on to study civil engineering at Stuttgart University. Leonhardt was an active and intrepid traveler in his youth, who would think nothing of walking from the northern rim of the Alps, down the length of Italy to Sorrento, in southern Italy, during his vacations. The beauty of the landscape, the magnificent mountain settings, and the unspoiled valleys were to have a lasting impression on him.

Much later, when he was a highly respected bridge engineer with his own practice, he would recall these memories in writing one of the great modern bridge books, entitled simply *Bridges*, in which he tackled the difficult subject of bridge aesthetics. It is a wonderful book, full of good ideas and sound advice and a bible on modern bridge aesthetics.

He won a scholarship as an exchange student in 1932 and went to Purdue University at Lafayette in central Indiana. It was a five-day sea journey on a White Star liner to New York. Having arrived in New York, Leonhardt lost no time in arranging to visit the two greatest living bridge engineers in America, David Steinman and the very elderly Othmar Amman. He already had the bug for bridges before he arrived in New York but this meeting with the great men, to discuss suspension-bridge design, was to inspire him for the rest of his life. As usual between semesters, Leonhardt was off walking, and this time he managed to hitchhike his way all around the USA and down to Mexico.

On his return to Germany he got a job with the government Autobahnen (freeways) to design and build road bridges. His boss, Karl Schaechterle, was later transferred to the Berlin department, with the mission to improve the aesthetic design of bridges with the help of the famous architect Paul Bonatz. Leonhardt followed his boss to Berlin and had a "fruitful time." While there he met and married Liselotte.

ABOVE: David Steinman, who is one of Leonhardt's engineering idols, standing on the Brooklyn Bridge.

LEFT: Fritz Leonhardt in 1993.

RIGHT: The Cologne suspension bridge with its slim plate girder deck (1938).

BELOW: Knee Bridge, one of the first cable stay bridges to be built in Germany (1956).

"She was a wonderful person who gave birth to our six children," he reflects. "She had to overcome the grief of losing two of our lovely daughters early on. Liselotte was never jealous of my love for bridges and provided a harmonious home, where I could always gain strength in difficult times."

The freeway network was expanding at an alarming rate trying to keep pace with the industrial growth of cities along the Rhine valley. Germany had never needed to build a suspension bridge, since the steel truss and concrete arch were sufficient for most river spans. But the freeway demanded long elevated approaches above street level and the shoreline, and this required more ambitious bridging structures.

Early in 1938 Leonhardt was given the responsibility of designing Germany's first suspension bridge across the Rhine near Cologne with a clear span of 1,240 feet. He took great care to design an elegant bridge with good proportions, which was detailed with care. "Even the rivets were placed in good order," said Leonhardt. The slim, plate-stiffened bridge deck that he built was a departure from the heavy truss girder of the American designs. This innovative deck was later copied by Amman for the beautiful Bronx Whitestone suspension bridge and the ill-fated Tacoma Narrows Bridge. However, the plate-stiffened bridge deck had unfortunate consequences for suspension bridges when the deck was high above the water line, unlike the Cologne bridge where it was low. The high winds can cause the deck to oscillate at an alarming amplitude, and actually caused the Tacoma Narrows to collapse.

World War II interrupted Leonhardt's bridge work, when he was sent to Estonia to build factories to extract oil out of the shale deposits. At the end of it, with his

BELOW: Cologne-Deutz Bridge over the Rhine, an elegant steel plate girder bridge (1946).

TOP: The Helgoland cable stay bridge in Norway (1991).

ABOVE: Leonhardt's aerodynamic bridge deck influenced many future bridge designs including the Kurushima Bridge in Japan, seen here under construction.

home destroyed in the Berlin blitz, he and his family moved into a small house in the Black Forest, and he started up a new practice in Stuttgart.

Owing to the devastation of the bombing, there was a lot to do. The City of Cologne asked Leonhardt to rebuild the Cologne-Deutz Bridge over the Rhine—when completed it was the first slender, steel-box-girder bridge in the world. Then in 1952 he was asked to design ten Rhine bridges and a further five over the Mosele, which had been destroyed in the war. The Rhine bridges were among the first cable-stay bridges in the world. The Kniebrücke with a clear span of 1,050 feet was the second of Leonhardt's family of three cable-stay bridges at Düsseldorf, but they were not the first to be built.

The smaller Stromsund Bridge in Sweden, designed by Franz Dischinger, another German engineer, was first, finishing in 1955, a year before. It is difficult to know who was the first to conceive of the cable-stay bridge idea, because the cable-suspension principle was developed by human beings thousands of years ago. Probably the bridge that most anticipated the modern cable-stay bridge was the small concrete Temple Aqueduct, built over the Guadalete River in Spain by Torroja in 1928.

But there was no doubt that Leonhardt was regarded as the leading light, the modern guru of the cable-stay bridge. It is a more efficient structure than a suspension bridge—although it cannot span quite as far—because no large anchorages are needed at each end and it uses fewer cables to support the span.

The Kniebrücke designed by Leonhardt was to influence cable-stay construction for many decades to follow, with its cables arranged in an uncomplicated parallel or harp configuration. "I would like to acknowledge the contribution of René Walter and Jorg Schlaich, who developed the multi-cable stayed bridge to perfection," says Leonhardt modestly. "They developed the bridge deck to get more and more slender. Schlaich's Evripos bridge in Greece has only a deck thickness of 450 mm [18 inches] with no edge beam and a span of 215m [705 feet]—fantastic."

Equally fantastic was Leonhardt's Helgoland Bridge in Norway, which had a slim deck and span of 1,390 feet. It was worth noting that Jorg Schlaich learned his trade working for Leonhardt in his practice in Stuttgart for a number of years before eventually going on to establish his own consultancy.

ABOVE: Aesthetics was an integral part of Leonhardt's design philosophy, exemplified here in the Kochertal Viaduct near Gieslingen, Germany (1979).

After the failure of the Tacoma Bridge, Leonhardt worked on the development of an aerodynamic bridge deck to reduce wind oscillation. Wind-tunnel tests carried out on Leonhardt's prototype at the National Physics Laboratory (NPL) in England proved it had good wind stability. The aerodynamic deck behaves just like the reverse of an aerofoil, creating a net downward pressure in high winds and with no oscillations. When the Severn Bridge was being designed, the consultants Freeman Fox and Partners got to know about the wind-tunnel tests at the NPL and adopted Leonhardt's aerodynamic deck for the bridge.

The Severn Bridge was the first suspension bridge in the world to use a slender steel-box-girder deck, and has changed the way all suspension and cable-stay bridge decks have been designed since. Leonhardt's contribution to this innovation cannot be underestimated. Only Japanese bridge engineers still adopt the deep-truss-girder bridge deck based on the American model, although very recently even they have moved over to the aerodynamic deck for the first time, on the Kurushima suspension bridge.

Fritz Leonhardt has worked in many other fields of engineering besides bridge design: his practice, Leonhardt Andra and Partners, designed the Munich Olympic

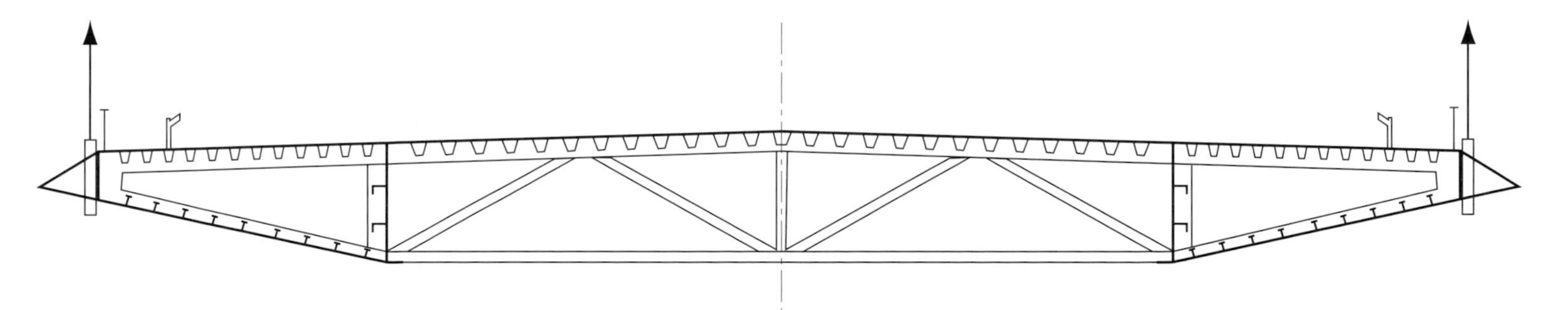

BELOW: The aerofoil deck section of the Kurushima suspension bridge.

ABOVE: Leonhardt's engineering versatility shown here in the concrete arch of the Maintal Rail Bridge in Veitshöchheim, Germany (1987).

Stadium cable-net roof canopy with the architect Frei Otto; he was the partner responsible for over 200 telecommunications towers, designed in both concrete and steel, which include the Stuttgart TV Tower, Hamburg, Mannheim, Cologne, and many others. He has designed bridges in Brazil, Venezuela, India, the USA, and Japan and has acted as technical adviser to governments on numerous major bridge schemes. He has been a professor of engineering at Stuttgart University for many years and was the rector there for two years between 1967 and 1969. He has received many international honors and decorations for his achievements, including an honorary doctorate from the University of Bath in England and the Gold Medal of the Institution of Structural Engineers in London.

Leonhardt, now approaching his 90s, will be best remembered for an epistle on bridge aesthetics he wrote in his book, *Bridges*— a discipline he feels has been neglected in modern bridge design.

> The esthetic qualities of the built environment have a profound influence on the human condition and the health of a society. Bridges are an essential part of the built environment. Massively, crudely shaped bridges, particularly freeway overhead bridges can destroy an environment, and create ghettos where crime and poor health flourish. Bridge esthetics should and must be the aim of every bridge that is designed or built in the world today. Artistic advice from suitably qualified architects on the design of bridges should not only be cultivated in the future, but also actively encouraged.

Jean Muller (b. 1925) "A resolute desire to simplify form"

That phrase was to stay in the subconscious of Jean Muller for the rest of his life, and to influence his approach to bridge design and civil engineering. It was spoken to him in 1952 by his boss, Eugene Freyssinet, a genius whose creative engineering skill has very obviously passed from master to pupil, when you consider the vast array of Jean Muller's accomplishments. There are the well-known and not so well-known pioneering bridge structures, barrel-vaulted dams in Africa and the Middle East, nuclear reactor plants in France, concrete-hulled supertankers, oil platforms, and harbor works—many of these structures were the first in their field; several were patented.

"His mind is like a volcano, and he has the energy of a twenty-five-year-old," says Jean Marc Tanis, chief executive of Jean Muller International (JMI). "He is unique and is still innovating, coming up with brilliant ideas on how to build or design a structure economically. Recently we asked his advice on a bridge we were designing for an international bridge competition in Malaysia and he came up with the most fantastic solution for a long-span cable-stay structure. We won the competition."

Born in Suresnes, a suburb of Paris, in 1925, Muller was one of five children and the middle one of the group. He followed his older brother Jacques into the construction industry and worked with him when he joined Campenon Bernard. His brother stayed on for 40 years, unlike Jean, who went on to form his own practice. His father worked in the largest bank in France, the CCF, in the foreign department, while his mother looked after the family and organized family gatherings and party games.

"My mother had the greatest influence on me, teaching me moral values, and social etiquette," Muller recalls fondly. "She was very strict with us and we could not go out late at night nor meet with girls. She encouraged my fondness for music and the piano. Playing the piano was a very rewarding pastime when I was growing up during the war years. I could play quite well."

At school he preferred physics and mathematics, and did not like the written subjects like literature and history. In 1944 he entered the Ecole Centrale des Arts des Manufactures for a four-year degree course in mechanical science, specializing in his final year in construction rather than the mining or chemical-engineering options. Immediately after graduating he went to work for STUP, the firm of Eugene Freyssinet. "The president of STUP was

ABOVE: Boulounais Bridge, France (1997).

LEFT: Jean Muller with a portrait of his mentor Eugene Freysinnet.

ABOVE: The flat steel arch of Nemours Bridge over the Grande Canal du Havre, Normandy, France (1995).

RIGHT: A typical precast concrete segmental section.

introduced to me while I was at university and encouraged me to take up construction in my final year. I had also heard a lot about Freyssinet and was very attracted to construction," says Muller.

He spent nearly five years with STUP, and worked on a number of civil-engineering projects, enjoying his time in Venezuela designing the Caracas Bridge, a large concrete-arch structure, and later on in New York, where he was sent to start up a new office.

It was while he was in New York promoting prestressed concrete that Muller developed the idea of match casting—precast box-girder sections.

"The Shelton Road Bridge just north of New York was not a huge span, but it was too long to precast and to transport to site in one piece. So I decided to cast the box-girder beam in three pieces. To make sure we had a precise fit on site, we match-cast the adjoining sections using the face of the first piece as the mold for casting the next pieces, rather like making a jelly from a jelly mold," Muller explains.

The sections were taken to site and assembled on a scaffold, where the butt joints were glued with epoxy resin and then stitched with prestressing cables threaded through the units. The beam was then stressed and, after grouting up the cable ducts, the scaffolding was removed. To this day it looks pretty good, with no signs of corrosion or spalling to be seen. Not bad for a bridge that was built in 1952.

This was the prototype for later match-cast, glued, segmental bridges designed by Muller and copied by bridge engineers the world over. Freyssinet had suggested this idea to him during a private conversation years before the Shelton Bridge, and as Muller recalls, "You don't forget a private discussion with Freyssinet: he does the thinking and talking while you listen."

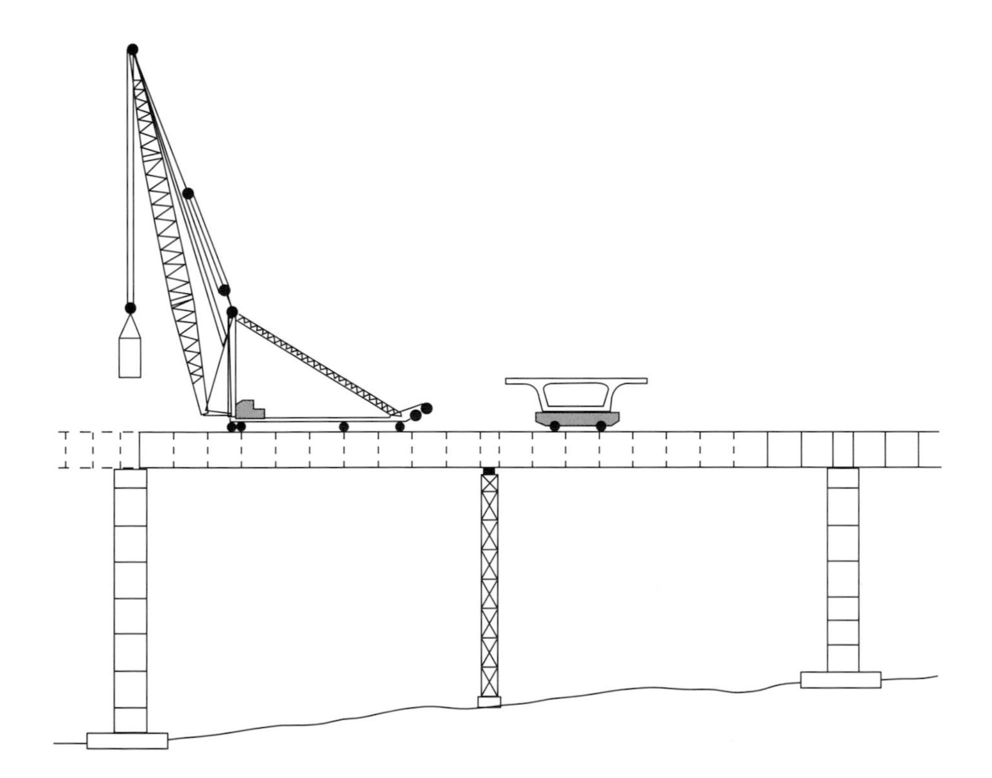

ABOVE: Ponchartrain Bridge under construction, New Orleans, USA.

LEFT: The erecting procedure for the Linn Cove viaduct.

ABOVE: The Confederation Bridge across the Northumberland Strait, Canada (1997).

TOP RIGHT: Linn Cove Viaduct, North Carolina, USA (1982).

RIGHT: Launching the box girder span of the Linn Cove viaduct.

Before he left the USA to return to France in 1955, Muller worked on the construction engineering for building the Louisiana Rail Bridge over Lake Pontchartrain, west of New Orleans. It was 23 miles long, had 2,170 spans of precast, prestressed concrete, and is the second longest bridge in the world.

He was made chief engineer of Campenon Bernard, then technical director, and finally promoted to scientific director of the group in 1976. During this time he was responsible for the development of bridge-building techniques using concrete segmental construction, and worked on oil platforms and even proposed a supertanker with a concrete hull during the Suez crisis. He is proud of the 426-foot-long, multispan Choisy-le-Roi Bridge over the Seine just upstream of Paris, built in 1962, with its glued-segmental, balanced-cantilever construction and long-line method of precast production—the first time this method was used anywhere in the world.

Muller also spent quite a lot of his time with Campenon Bernard designing large multi-arch dams. In Algeria at one stage when he was designing the Erraguene Dam he was caught up in the a civil war when the Feligas rebels tried to overthrow the government. "I kept my head down and tried to keep my distance from the bullets and mortar bombs that were being fired in the capital."

TOP: Lowering a box girder segment into position for the Linn Cove viaduct.

ABOVE: Choisy-le-Roi Bridge over the Seine (1962).

OPPOSITE ABOVE: Brotonne Bridge near Rouen, France (1977).

OPPOSITE BELOW: The Sunshine Skyway, Tampa, Florida, USA (1986).

BELOW: Sallingsund Bridge, Denmark (1974).

Of those that he designed he remembers vividly the Latyian Dam in Iran close to Tehran, with the beautiful snow-capped mountains in the background; Sourlages Dam in France with its barrel-vaulted arch and two great buttresses; the Sefidad Dam in Iran, a huge gravity dam 360 feet high, which needed over 1 million cubic meters (28,250,000 cubic feet) of concrete; but the best of all was the Djin Djin Dam in Algeria, built among the ruins of the Roman Empire. Then there was a period of designing prestressed-concrete containment structures for nuclear reactors when Europe went nuclear for new energy sources. The Tiage reactor in France was the first to use heavy water as the moderator, with natural uranium as the catalyst. It was a cleaner and safer system than Westinghouse's enriched-uranium rod process, but it never caught on outside France. Muller's team designed the prestressed concrete pressure vessel for the nuclear reactor.

When the demand for nuclear power stations dwindled, Muller went back to the USA to see if he could secure bridge contracts to keep the technical team at Campenon Bernard fully employed. A chance meeting with Eugene Figg in 1978 led to an agreement to combine resources to tender for the really big bridge projects that were coming up in the South. And thus Figg and Muller Engineers was set up, jointly owned by Campenon Bernard, Muller, and Figg.

They took the US bridge market by storm. The long, multispan Florida Keys bridge was followed by the ingenious Linn Cove Viaduct cutting a swathe through the forested slopes of the North Carolina mountains. Muller devised a span-by-span, gantry-launched assembly of the precast box-girder beams, and the building of the next pier support from the finished span in order to cause the minimum disturbance to the fauna and flora of this beautiful region. "Where possible we tried to save as many trees as possible: only those on the direct line of the viaduct had to be cut, but no more, as we eliminated the need for site-access roads during span erection by using the completed span itself," he recalls.

When Linn Cove was finished it seemed as though the bridge had been dropped into place from the sky. There were very few scars and telltale spots to show where any excavation, temporary works, and construction traffic had been.

Then, in 1986, came the landmark bridge of the state of Florida and one of the greatest modern bridges, the cable-stayed "Sunshine Skyway." It's an attraction in its own right. "I designed the concept myself, the pronounced slope of the center span, the shape of the towers, the cable fans arrangement and so on ... There were no architects," says Muller, quite modestly.

After this great success Figg and Muller disbanded, and Jean Muller went on to establish JMI, Jean Muller International, a world-class bridge design consultancy with offices in San Diego, Tallahassee, Paris, and Caracas. Under the JMI flagship emerged the highly acclaimed, cable-stayed Isère Bridge in Romans, regarded as one the Europe's finest bridges; the externally prestressed-concrete

Long Keys Bridge, a decisive breakthrough in new technology, which was the precursor to many similar bridge spans throughout the world; and further refinement of gantry-launched and segmental construction with the H3 viaduct in Hawaii and then the Prince Edward Island Bridge across the ice floes of Canada.

Eugene Freyssinet was already using prestressed-concrete segments for constructing the famous Marne bridges in the late 1940s when Jean Muller joined his practice. Since then, Muller has continued to promote this process to its current state in the art. He evolved match-cast segmental joints, the first launching gantries, and external prestressing for box-girder segments, and was the first to use precast segmental construction for a cable-stay structure on the Brotonne Bridge. But that is not all: very recently at the age of 64 he schemed the preliminary design for a massive long-span, cable-stay bridge in Malaysia—it is yet another breakthrough in cable-stay construction, allowing spans of around 3,300 feet to be built.

"It is more efficient than the record-breaking Pont de Normandie," says Muller, "because it needs only one plane of cable strands, rather than two, and incorporates a stiff but lightweight composite deck over the central span, where the top slab section is prestressed concrete, but the section below is a tubular-steel truss." It will be the longest cable-stay bridge in the world when it is built.

A bridge in the hands of a genius like Jean Muller is not just a utilitarian structure: it becomes a work of engineering art.

OPPOSITE ABOVE: Chillon Bridge, Switzerland (1966).

OPPOSITE BELOW: Pont Isère, Romans, France (1991).

LEFT: CAD image of the record breaking Sungai Johore, also known as "The Ceremonial Bridge" in Malaysia, now under construction.

BELOW: Roize Bridge, Grenoble, France (1992).

Glossary

Abutment: the end support of an arch bridge, built on the bank, which resists the horizontal thrust of the arch(es).

Aerodynamic deck: a bridge deck with an aerofoil cross-section which provides aerodynamic stability in high winds.

Air spinning: a method of making the cables for a suspension bridge. The individual wires are passed back and forth across the entire span, until the suspension cable is built up to the required diameter.

Anchorage: the support structure for anchoring and holding the cable ends of a suspension bridge.

Arch: a curved bridge span where the weight and forces acting on the span are carried in compression outward to the supports.

Aqueduct: a bridge or viaduct carrying a canal or channel of water across the span(s).

Barrel Vault: an arch of semicircular shape, often of brickwork or masonry, whose length is longer than its span. Sometimes called a cylindrical vault.

Bascule: a type of drawbridge with a counterbalance weight at the ends, which causes the span to rise and open when it is engaged.

Beam: a narrow and deep, or shallow and wide, rectangular member or an I-section or T-section structural element, spanning between pier supports. A bridge is often made up of number of these beams. Also known as girders when they are deep.

Bedrock: the solid rock layer beneath the silt and sand of a river bed or estuary.

Bends: see **Caisson disease**.

Bowstring arch: an arch whose ends are linked or tied together to resist the outward thrust.

Box girder: a deep hollow box beam which can have a rectangular or trapezoidal cross-section.

Cable-stayed bridge: a bridge whose deck is supported directly by a series of inclined cables connected to the pylon or mast and not from the vertical hangers of a suspension bridge.

Caisson: a structure for keeping water out while deep foundations are being excavated. In the lower working chamber water is kept out by increasing the air pressure.

Caisson disease: a disease that can affect workers working under compressed air who come too quickly out of the airlock. It is caused by bubbles of nitrogen coming out of the blood. Also called decompression sickness or the bends.

Cantilever: a structure or beam that is unsupported and free at one end, and fixed and supported at the other.

Cast in place: concrete that is formed in its final position by placing wet concrete into formwork or shuttering and allowing it to harden: it is the same as *in-situ* concrete.

Cast iron: a brittle alloy of iron with a high carbon content that is good in compression and weak in tension.

Catenary: the sag or profile of a rope or cable suspended from two points, as on a suspension bridge.

Cement: a powder that, when mixed with water, binds a stone-and-sand mixture into a strong concrete within a few days.

Centering: a temporary framework of timber to support the masonry while an arch is being built.

Chord: the top or bottom horizontal part of a truss.

Cofferdam: a structure for keeping out water to allow excavations and building of foundations in ground below water level. It differs from a **caisson** (q.v.) because it is open to the air.

Compression: a force that tends to shorten and compress something.

Concrete: a mixture of sand, stone and water bound by **cement** (q.v.), which hardens into a rocklike material.

Corbeling: successive layers of masonry or brick projecting beyond each other.

Creep: the slow permanent deformation of a material under stress, a characteristic of concrete.

Crown: the highest point of an arch.

Cutwater: the end of a pier base that is pointed to cleave the water.

Deck: generally the top side of a beam, box girder, or truss which forms the running surface for vehicles or pedestrians.

Decompression sickness: see **Caisson disease**.

Elliptical arch: an arch with a curve that becomes much tighter toward the crown.

Empirical formula: a formula or design rule based on one or many series of observations but with no theoretical backing.

Eyebar: the unit from which the chains of a suspension bridge were constructed, with a flattened ring at each end for linkages.

Falsework: temporary support scaffolding used during construction.

Fan configuration: the arrangement of cables on a cable-stay bridge which fan out from the pylon just like the folds of a paper fan do when it is open.

Flange: the top and bottom plates of a box girder, plate girder, or an **I-beam** (q.v.).

Formwork: temporary timber or metal shuttering to contain and support concrete while it hardens.

Girder: a large or deep **beam** (q.v.).

Hangers: the wires or bars from which the deck is hung from the cables in a **suspension bridge** (q.v.)—also known as suspenders.

Harp configuration: a cable-stay arrangement where the stays radiate out parallel to one another like the strings of a harp.

Haunch: part of the arch between the springing and the crown.

I-beam: a beam or girder with an I-shaped cross-section.

***In-situ* concrete:** see **Cast in place**.

Keystone: the voussoir at the crown of the arch.

Navigation span: the part of a bridge with maximum clearance for shipping.

Ogival arch: a pointed arch.

Pier: the support of a bridge deck span, that is not on the bank. It is also a general term used for the base or foundation of a bridge.

Plate girder: a flat bridge deck with a shallow rectangular section.

Pointed arch: an arch with an angle at its crown.

Pontoon bridge: a bridge formed from boats, logs, or drums floating on the water and tied together.

Portal: a frame with uprights connected by a horizontal member at the top, just like the goal posts in soccer.

Precast concrete: concrete that has been hardened, taken out of its formwork or shuttering, then transported to the site and placed in position.

Prestressed concrete: steel wires or strands within the wet concrete are stretched (pretensioned) and held fast until the concrete has hardened. The tension in the wires keeps the concrete in compression and gives it a greater tensile strength. Alternatively, the strands are threaded through plastic or metal tubes in the hardened concrete and after the strands are tensioned (post-tensioning) they are anchored and grouted into place.

Pylon: the vertical mast or tower above the bridge deck to which the cable stays are fixed.

Reinforced concrete: concrete that is strengthened with high-tensile steel bars or rods to give the concrete tensile strength and ductility.

Segmental arch: arch formed from the segment of a circle.
Semicircular arch: arch forming a half-circle.

Shrinkage: the shortening of concrete that occurs as it dries and hardens.

Side span: the outer or end spans of a suspension bridge, from the tower to the anchorage, balancing the central suspended span.

Side sway: the side-to-side movement of a bridge deck in a wind.

Spandrel: the area of an arch above the **voussoirs** (q.v.) and below the bridge deck.

Springing: the point where the end of an arch meets an abutment or a pier.

Steel: an alloy of iron which has more carbon than **wrought iron** (q.v.) and less than **cast iron** (q.v.), combining the tensile strength of one and the compressive strength of the other.

Stiffening truss: a truss usually beneath the entire deck of a **suspension bridge** (q.v.).

Striking: the action of removing formwork in concrete and centering from beneath a completed arch.

Suspension bridge: a bridge with its deck supported by large cables or chains draped from towers.

Tensile strength: the ability of a material to withstand tension.

Tension: a force that tries to stretch and lengthen something.

Torsion: the stress produced when a structure is being twisted

Tower: the vertical support structure of a suspension bridge from which the cables are hung.

Truss: a frame of tension and compression members which together make up a long-span beam.

Viaduct: a bridge carrying a road or a railroad.

Voussoirs: the wedge-shaped stones from which an arch is formed.

Web: the side plates of a box girder or the vertical plate of an I-beam.

Wrought iron: iron that has been hammered into shape, with a low carbon content, low compressive strength, and high tensile strength.

Zigzag suspension: the arrangement of suspension hangers in zigzag fashion rather than vertical for added wind stability, and first introduced on the Severn Bridge, which links England with Wales.

Bridge facts

Longest bridge in the world

The longest bridge in the world is a viaduct 90 miles long in the Hwang Ho Valley in China, followed by the Lake Pontchartrain trestle bridge near New Orleans, which is 23 miles long. In Europe the longest bridge over water for many years was the Oosterschelde in Holland with a length of 3.4 miles. It has recently been beaten by the Oresund crossing linking Denmark to Sweden, with a bridge length totaling 8.3 miles, and which includes the record-breaking suspension span of the East Bridge.

Largest bridge-building program

The largest civil-engineering undertaking and the most expensive and colossal bridge-building program ever planned is taking place in Japan as I write this. There are no fewer than 17 major bridges being built—each one in itself would be a major undertaking for any other country in the world, but not Japan. This multibillion-dollar program will run for 15 years, and will see some of the longest and tallest cable-stay and suspension bridges ever built. The bridges will link the mainland of Honshu to the island of Shikoku via three major arterial highways—two of which carry railroads—across the inland sea of Seto. To the south will be the Onomichi Imabari highway, with a total of nine bridges, one of which includes the Tartara bridge, the longest cable-stay span in the world, and the three elegant Kurushima suspension bridges. To the north is the comparatively short Kobe Naruto highway, which has only two big bridges, but one of them is the Akashi Kaikyo, the longest span in the world. In between these two highways is the Kojima Sakaide route with a mixture of cable-stay and suspension bridges, totaling six in all.

The record for the longest bridge span

Name	Year	Location	Type	Span in feet
Trajan's Bridge	104 AD	Danube River	timber arch on stone piers	170
Trezzo	1371	Italy	stone arch	236
Wettingen	1758	Germany	timber arch	390
Menai Straits	1826	Wales	chain suspension	580
Fribourg	1834	Switzerland	wire-cable suspension	870
Wheeling	1849	West Virginia	suspension	1,010
Lewiston	1851	Niagara Falls	suspension	1,043
Cincinnati	1867	Ohio River	suspension	1,057
Clifton	1869	Niagara Falls	suspension	1,269
Brooklyn	1883	New York	suspension	1,595
Firth of Forth	1889	Scotland	cantilever	1,700
Quebec	1917	Canada	cantilever	1,800
Ambassador	1929	Detroit	suspension	1,850
George Washington	1931	New York	suspension	3,500
Golden Gate	1937	San Francisco	suspension	4,200
Verazzano	1965	New York	suspension	4,260
Humber	1981	England	suspension	4,624
East Bridge	1998	Denmark	suspension	5,328
Akashi Kaikyo	1998	Japan	suspension	6,529

Bridge type and longest spans

Type	Span in feet	Year	Location
Cantilever truss	1,800	1917	Quebec, Canada
	1,710	1890	Firth of Forth, Scotland
	1,644	1974	Commodore John Barry, Delaware, USA
	1,500	1943	Howrah, Calcutta, India
Steel arch	1,700	1981	New River Gorge, West Virginia, USA
	1,675	1931	Bayonne, New Jersey, USA
	1,670	1932	Sydney Harbour, Australia
Concrete arch	1,280	1979	KRK, Zagreb, Yugoslavia
	1,000	1964	Gladesville, Sydney, Australia
	951	1964	Foz-do-Iguaco, Brazil
Steel box girder	984	1974	Rio de Janeiro, Brazil
	837	1956	Sava, Yugoslavia
	831	1966	Zoo, Cologne, Germany
Cable stay	2,919	1998	Tartara, Hiroshima, Japan
	2,808	1995	Normandie, France
	1,984	1996	Quingzhou Minjang, China
	1,974	1993	Yangpu, China
Suspension	6,529	1998	Akashi Kaikyo, Japan
	5,328	1998	East Bridge, Denmark
	4,625	1981	Humber Bridge, Hull, England
	4,260	1964	Verazzano, New York, USA
	4,200	1937	Golden Gate, San Francisco, USA

Useful references

Brown, David: *Bridges—Five Thousand Years of Defying Nature*, Mitchell Beazely, UK, 1993

Delony, Eric: *Landmark American Bridges*, ASCE, USA, 1993

Gies, Joseph: *Bridges & Men*, Cassell & Co, UK, 1963

Hayden, Martin: *The Book of Bridges*, Marshall Cavendish, UK, 1976

Kingston, Jeremy: *How it is made: Bridges*, Threshold Books, UK, 1985

Leonhardt, Fritz: *Brücken (Bridges)*, the Architectural Press, UK, 1982

Plowden, David: *Bridges—The Spans of North America*, Viking Press, USA, 1974

Salvadori, Mario: *Building—from caves to skyscrapers*, Atheneum, USA, 1979

Steinman and Watson: *Bridges and Their Builders*, G.P. Putnams, USA, 1941

Wittfoht, Hans: *Building Bridges*, Beton Verlag, Germany, 1984

Useful Sources of Information

- Caltrans, District 4 Public Affairs, 111 Grand Avenue, Oakland, CA 94612, USA
- Ecole Nationale Des Ponts et Chaussées, Centre de Documentation, 6–8 avenue Blaise Pascal, Cité Descartes, Champs sur Marne, 77455 Marne La Vallée, Cedex 02, France
- HAER, US Department of the Interior, National Park Service, 1849 C Street, NW, Washington, DC, 20240, USA
- ICE Library, The Institution of Civil Engineers, 1 Great George Street, Westminster, London SW1P 3AA, UK
- MTA, Public Affairs, Bridges & Tunnels, 18th Floor, 10 Columbus Circle, New York, NY 10019–1203, USA
- New York Port Authority, Customer Relations, Tunnels, Bridges & Terminals, One World Trade Center, New York, NY 10048, USA

Index

D0926639

February 24, 1836 September 29, 1910

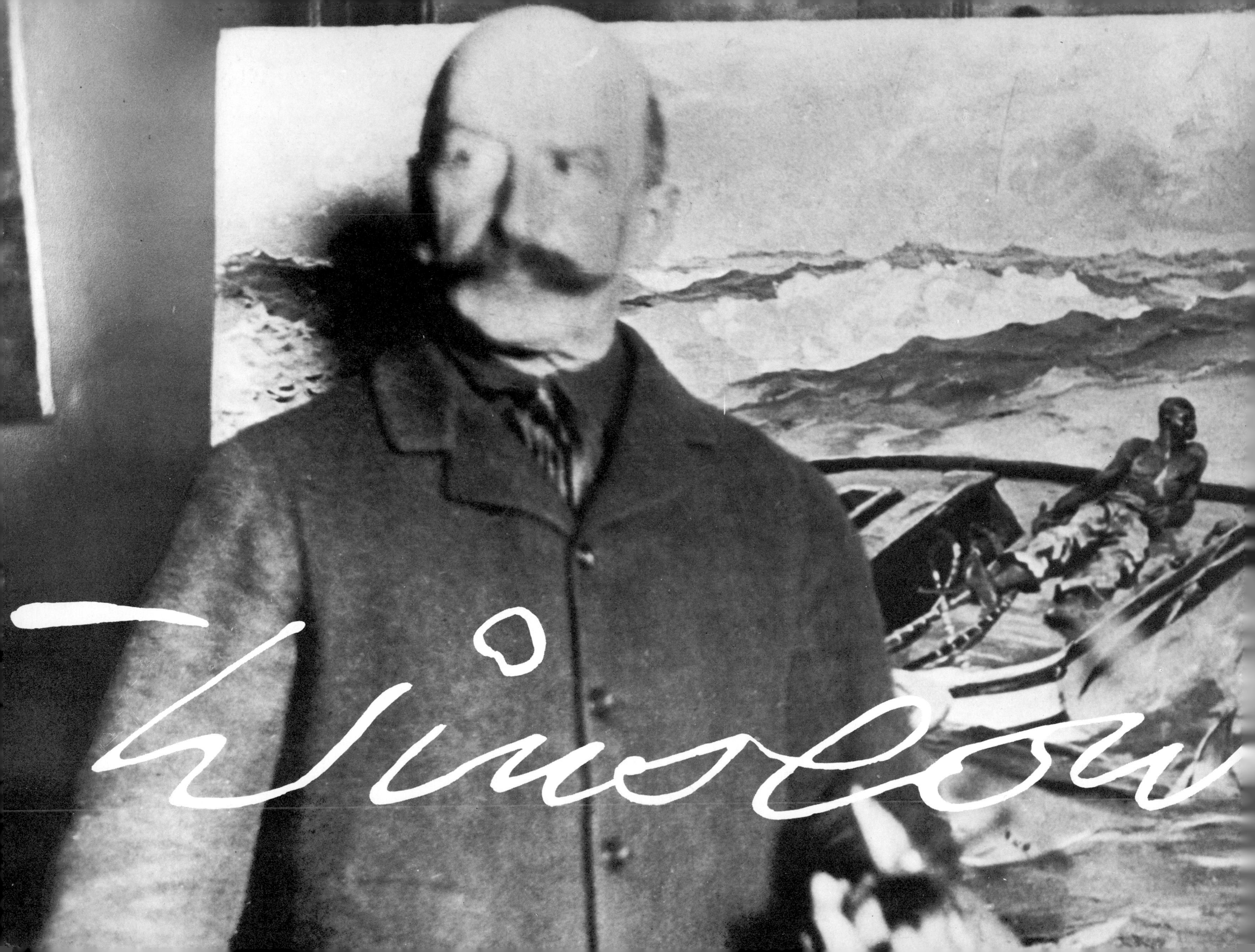
Winslow

Frontispiece:

This is the only known photograph of Homer while working on his famous oil, *The Gulf Stream,* in his Prout's Neck studio. Courtesy of the Homer Collection, Bowdoin College Museum of Art, Brunswick, Maine.

WINSLOW HOMER IN THE TROPICS

By Patti Hannaway

Foreword by Dr. Hereward Lester Cooke
Curator of Painting at the National Gallery of Art
Washington, D.C.

Richmond, Virginia
A Media General Publication

Published in the United States of America
by Westover Publishing Company
333 East Grace Street, Richmond, Virginia 23219

Printed in Japan by
Dai Nippon Printing Co., Ltd.
Library of Congress Catalog Card Number: 73-79127

ISBN Number: 0-87858-038-7

For my husband, Walt Hannaway, whose unfaltering faith and incessant prodding provided me the much needed impetus to complete a fervent dream.

In acknowledgement

Much gratitude is due to the many previous biographers of Winslow Homer as well as to curators and researchers at various art museums who have willingly provided assistance in compiling data on the artist's tropicals. I am especially indebted to Philip C. Beam, curator of the Homer Collection, Bowdoin College, Brunswick, Me., for his advice and continuing encouragement. Jean Gould, author of an earlier portrait of the artist, also afforded clarification in a number of special areas; Lloyd Goodrich, advisory director for the Whitney Museum of American Art, and his wife, Edith, were helpful in offering details as to ownership of some missing works; Charles Homer Willauer, the late artist's great nephew, graciously provided the 1902 sketch of his "Uncle Win" drawn by Marguerite Downing Savage. Frank J. Doherty, New York, through the Bacardi Corporation, put me in contact with Señor Francisco de Pando, Madrid, Spain, son of the late subject appearing in the *Governor's Wife.*

The development of certain details was possible via counsel with Susan B. Carmalt, former curatorial assistant at the Museum of Art, Rhode Island School of Design, Providence; Don McCue, U.S. Consulate General to Bermuda; Mrs. Terry Tucker, historical research department, Bermuda Library, and Elizabeth H. Larkin, consultant with Keppie Travel Bureau, Clearwater, Fl. Counterpoint data and assistance was provided by Mary MacRae, Homosassa, Florida; Miss S. Loeb and Herbert G. Hamilton, Museum of Fine Arts, Boston; Hugh J. Gourley, III, director, Colby College Art Museum, Waterville, Maine; James Mahey, director, Cummer Gallery, Jacksonville, Florida; Evan H. Turner, director, Philadelphia Museum of Art; Roy Craven, director, University Gallery, University of Florida, Gainesville; James K. Kettlewell, The Hyde Collection, Glens Falls, NY; Edward W. Lipoqicz, Canajoharie Library & Gallery, Canajoharie, NY; James L. Enyeart, University of Kansas Museum of Art, Lawrence, Kansas; Ellen W. Lee, Indianapolis Museum of Art.

Further cooperation was received from Margaret Olson, Minneapolis Institute of Arts; Virginia Morse, Fine Arts Gallery of San Diego; Carla M. Martin, Cincinnati Art Museum; Billie Sass, Toledo Museum of Art; Gloria Sullivan, IBM Corporation; Sarah Blank, Pennsylvania Academy of Fine Arts; Elaine E. Dee, The Cooper-Hewitt Museum of Decorative Arts & Design, Smithsonian Institute; Truth W. Crooks, The Worcester Art Museum; Betty R. Saxon and Allen Wardell, Art Institute of Chicago; Linda Ferber and Arno Jakobson, The Brooklyn Museum; Linn Orear and Mary Mancini, Fogg Art Museum; Cynthia Carter and Marceline McKee, The Metropolitan Museum of Art.

My appreciation to The Viking Press, Inc., New York, for allowing me to quote from *The Story of Modern Art* by Sheldon Cheney, Copyright 1941, 1958 and 1969; also to the American Heritage Publishing Company, New York, for allowing me to quote from *Horizon* a passage of John Canaday's.

Helpful reference materials were provided by Shirley Thayer, Maine State Library; Theresa Cederholm, Boston Public Library; Mary O. McRory, Florida State Library; John L. Lochhead, The Mariners Museum, Newport News, Va.; Mrs. T. O. Bruce, Monroe County Public Library, Key West; Theodore Leslie, Hillsborough County Historical Commission, Tampa; Margo Williams, Sawyer Free Library, Gloucester, Ma.; E. H. Lorenzen, Public Library, Townsend, Ma.; Vicar Ernest B. Pugh, All Saints Episcopal Church, Enterprise, Fl.; and Ernest C. Myer, former principal of Hurley School, Hurley, NY.

A hearty word of thanks goes also to private collectors of Homer's tropicals who cooperated by allowing use of reproductions in this volume: Mrs. Doris M. Brixey, New York; Harry L. Dalton, Charlotte, NC.; Mrs. Charles Henschell, New York; Arturo Peralta-Ramos, New York; Laurance S. Rockefeller, New York; Mortimer Spiller, Buffalo, NY.; Charles Woodward, Philadelphia, and those galleries such as Hirschl & Adler, Kennedy Galleries and Wildenstein & Company which assisted most graciously.

Sincere appreciation is also expressed to Ed Harrell, Sam Strickland, Beverly Pinette and Tanyua Franz of Westover Publishing Company for their thoughtful assistance and unstinting consideration during this manuscript's publication processes.

Without the cooperation of all of these persons, this view of Homer could not have been presented.

P. H.
Clearwater, Florida

Contents

Foreword

Homer is one of the very few American artists whose standing in the art world has never wavered. He was elected to the National Academy of Design while still in his twenties. He represented his country abroad at international expositions while still a young man, and he sold a painting, *The Gulf Stream,* for the highest price ever paid to a living American painter.

After his death his status did not ebb or falter; on the contrary his fame and prices continued to rise steadily. Nor have his paintings languished in stylistic purgatory, waiting for art historians on their self-appointed rounds of sifting the chaff from the wheat.

In spite of the fact that he ranked among the most successful artists in America for most of his career, we can surmise from oblique references in his letters that he did not make enough money to live on from the sale of his paintings. Details of the Homer family finances will probably never be in the public domain. However, it seems most of Homer's income came from shrewd real estate deals and rents paid by summer visitors at Prout's Neck. In spite of a lifetime of triumph, Homer never considered that he had achieved his goal. Instinctively he knew that it was his watercolors which would ultimately be considered his most important contribution, and he was right. His big oil paintings of the sea, in all their power and majesty, are landmarks in American painting, but his watercolors opened up a new chapter in the history of art. His exploitation of the transparency of the medium, the calligraphic quality of his brushwork, and the spontaneity and unexpected effects he made possible with a fluid and often unpredictable medium, opened new horizons for the art of watercolor painting, and set it apart from its senior partner, oil painting. Only with an instant technique could the transitory effects of nature, particularly the scintillating effects of tropical sunlight be captured. Homer instinctively knew that this was his forte and the most natural expression of his talent. "You will see," he was quoted as saying, "I will live by my watercolors." History has endorsed this prophecy.

Homer's biographers have in general stressed his rugged individualism, his quest of solitude, his insistence on secrecy, so that one wonders if he was trying to hide some dark secret. From his letters one occasionally gets the idea that he painted only when the fish were not biting. From Mrs. Hannaway's text he emerges as a more human person, who had to make the agonizing decision of either marrying the woman he loved, or following the lonely path of his art; who was devoted to his dog,

Sam, and who was regarded by his taciturn neighbors in Maine as one of the most kind-hearted men who ever lived.

Another fact which emerges from Mrs. Hannaway's research is the close bond of deep affection which existed between the artist and his mother. An artist in her own right, she undoubtedly was the first influence in his life to recognize his talent and the singleness of his purpose and do what she could to set him on the road to future greatness.

The tropics, to those of Homer's generation, were outposts of paradise where beautiful people lived amid stage settings of moonlit beaches and exotic plants. This was the never-never land Henri Rousseau tried to conjure up, Joseph Conrad immortalized in prose, and Paul Gauguin found. Today the tropics have overtones of Guadalcanal, jungle rot and rusting machines of war, but at the turn of the century it was sunlight, beauty, and color. Homer has preserved not only these qualities but the feeling of a distant, happier world.

Dr. Hereward Lester Cooke
Curator of Painting
The National Gallery of Art
Washington, D.C.

This portrait of Winslow Homer (considered to be the best of him as he looked late in life) was created by Marguerite Downing Savage in 1902.

Mrs. Savage was twenty when she met the artist. Philip C. Beam in his biography, *Winslow Homer at Prout's Neck,* records that she ran out of yellow ochre, while painting at Prout's Neck. Ignoring "warnings about his unfriendliness," she knocked at Homer's door. "To her delighted surprise, he gave her a large tube of ochre in his favorite Winsor & Newton brand, and they became very friendly."

Showing great interest in her work, he "gave her many hints, and went out of his way to introduce her to people who might further her career. Once he brought John W. Beatty, then art director of the Carnegie Institute, to her studio to see her paintings."

Later, during a tea party at the *Ark,* "he asked her to do a drawing of him, and after several sittings it was completed." He apparently liked the portrait very much for "when Mrs. Savage signed and dated it on September 6, 1902, Homer paid her three, twenty-dollar gold pieces for it." It is the only one he ever commissioned another artist to do of him.

Introduction

Americans instinctively take to Winslow Homer. They see a clarity in his works that shows them their world as they like to think they see it too. They view New England, the Atlantic and, most of all, the tropics much as Homer did. It is especially Homer's tropics they love best, for through him they see a vision of freedom, a spirit of warm adventure linking the best of land and the best of sea. In the American dream, these places are just beyond the horizon and today just as much as when he first painted them, Homer's tropics beckon as strongly as ever.

With the ready access of air travel, the tropics are only hours away. The mystical charm of Bahamian isles and Florida hamlets are available for many to enjoy inexpensively, and in fact many hundreds of thousands have found their way there. And strangely, because it is within reach, that sense alone brings the wealth of Homer's art treasurers so much closer. These Americans see the tropics through the skilled eye and brush of Winslow Homer.

For Homer, the tropics was discovery. It took him long days on a sea voyage to reach—a voyage which he probably relished being a descendant of seafarers himself. It was an emerging world when he first set his eyes on the lush island foliage and dense Florida jungles. The accommodations were still rudimentary for even the wealthiest traveler and Homer personally disdained first class hotels for simpler haunts.

This volume sets out to show Homer's best arena, the tropics, and to tell something of how he acquired his love for the Gulf Stream islands and Florida, and to present many of the works he created there. For it is these paintings and watercolors which elevated him beyond the ranks of being just a good artist, to one of America's greatest.

The artist himself would certainly eschew any such appellation. He was by nature a solitary, withdrawn, shy man. In his later years he lived for his work and shunned the approbation his countrymen sought to shower on him. In the six decades since his death, countless volumes have detailed his idiosyncrasies and have purported insight into his aloof nature. But in summary, Homer was a unique man in many respects. He selected his lifestyle and pursued his craft as he saw it, allowing for only influences he personally chose to accept from his family, close friends and fellow artists. He was indeed, a rare individualist.

1

Family and Environment

Born February 24, 1836, in Boston, he was the second son of Henrietta Maria Benson Homer and Charles Savage Homer. Except for being New Englanders, there seems little else common to their backgrounds. She came originally from the down east Maine community of Bucksport which was named for her mother's father. His father's ancestry could be traced back to an English shipowner who sailed into Boston harbor in the mid-17th century. Henrietta was a witty, talented watercolorist, who took up painting as a pastime, and whose works were good enough to win exhibitions in professional shows. Gentle mannered and gracious, she contributed a truly feminine outlook to what eventually became a basically male household. Her firm hand maintained a close-knit family circle during her lifetime.

Charles Homer, a brusque, gregarious, all assuming man of impressive stature, was the proprietor of a hardware business. He continually devised outlandish, and usually wide of the mark, get-rich-quick schemes. His brother, James, as captain of his own ship, sailed throughout the West Indies and brought home tall tales of adventure and fortune. Charles, too, sought a taste of that sort of life at times.

Winslow Homer was born into a warm family oriented household—not overly elegant—but one of comfort filled with a great amount of love. Love for one another, for life and for nature—influences which would play vital roles in the life of a man destined to be a great artist.

2

Early Boyhood Years

Winslow Homer was the middle of three sons. He followed his older brother, Charles, Jr., by two years. As a toddler, he became fascinated with watching his mother paint and followed her every brush stroke with keen eyes. He was given a set of crayons at the age of three or four and used these to eagerly color the illustrations in the worldwide variety of periodicals which came regularly to the Homer residence.

Arthur Benson Homer was born in 1841 and shortly thereafter the family moved to Cambridge, Mass., a move prompted by the parent's desire to obtain for their sons the best education possible.

The Homer boys must have experienced a delightful childhood. Young Arthur was often, however, the scapegoat for his brothers' adventures. There were numerous Benson cousins nearby to add to the revelry. Theirs was the joyous freedom of the great outdoors. And, like boys the world over, Winslow and his brothers loved to run and play games in the wide meadows and catch fish in the small ponds near their Garden Street residence. Winslow never forgot these experiences.

In 1847, at the age of eleven, Winslow recorded the boisterous play of those happy, carefree days with precise pencil drawings. He depicted the exhilarating game of "the beetle and wedge" and titled it *The Youth of C.S.H., Jr.* (to honor his older brother). A small inset sketch carefully keyed the boyish figures appearing in the drawing. In another sketch, *Adolescence,* he showed a boy lazily daydreaming on a hillside.

He retained a number of his early boyhood drawings for some 30-years and drew on their content for later lighthearted nostalgic lithographs and watercolors. Those rural sports dear to the hearts of boyhood are found in his lithographs and watercolors created in the 1870's at Gloucester Harbor, Mass. and Hurley, New York.

Homer's adroit, lightning fast sketching ability, his keen eye and prolonged fresh feeling of youth became important factors which aided him greatly in his adult life. His sharp perception and imaginative vision, enabled him to render meaningful compositions quickly. Later, only minor refinements made in seclusion were necessary.

Winslow was not an avid or scholarly student. His fourth grade teacher once discovered him gazing out of the window sketching instead of doing his math. As punishment, he was told to stand in the dunce corner, remain inside the classroom during noon recess and memorize his arithmetic lesson. When he failed to comply with the last element of her punishment, the teacher sent a discouraging note home to Winslow's father.

Both parents evidentally understood his preference for drawing, since young "Win" escaped with only a minor threat from his father. Later, when visiting Europe, Mr. Homer sent him a set of lithographs by Bernard-Romain Julien. These were drawings of hands, ears, eyes, noses and faces, for the boy to study and copy.

Life was not always sunshine and happiness for the Homer family. When the gold rush of 1849 struck, Charles Savage Homer, Sr. was among those Bostonians who caught the fever. He sold his prosperous hardware firm to a former associate in order to raise sufficient cash to head for the gold fields. Using the proceeds from his sale, he sent a sailing vessel loaded with mining equipment around South America while he took a shorter route across country, through the Isthmus of Panama and north to California. Henrietta, trusting her husband's judgment, did not question the venture. But the enterprise was fraught with disappointment and delay—someone had jumped his company's claim. His efforts to re-establish the claim and assume control were costly and failed. It was two years before he returned to Cambridge and when he finally arrived, his patient wife had to pay the buggy fare.

Though nearly penniless, the family had managed to maintain the home on Garden Street. Mr. Homer by borrowing from relatives, found enough capital to again set up his own hardware business.

Young Winslow finished his secondary schooling in June of 1854. Charlie, Jr. had already begun his studies in chemistry at Lawrence Scientific School in Cambridge and was doing well academically. Since his middle son did not show any aptitude or inclination toward college, Mr. Homer began looking around for a job that Winslow might suitably fill. He approached several acquaintances in haberdashery but no one at that time needed a clerk with Winslow's dapper appearance. He had to seek employment elsewhere.

A week to 10 days later, Mr. Homer found an ad in the *Boston Star* under "Help Wanted." The brief advertisement read: "Boy wanted; apply to Bufford, Lithographer; must have a taste for drawing; no other wanted." The proprietor, John H. Bufford, was a member of the Cambridge Volunteer Fire Department, which had elected Mr. Homer foreman. An interview was easily arranged. With his mother's assistance, Winslow selected some of his best sketching examples and went into Boston on the bus with his father to meet Mr. Bufford.

Most applicants, as apprentices learning lithography, would have been obliged to pay $300 for the privilege. Impressed with young Homer's ability, Bufford reduced the fee to $100. Mr. Homer considered this a generous bonus. Henrietta was also pleased but more satisfied that her son was being launched in a good profession, one that had possibilities for advancement. She perhaps hoped, too, that a firm foundation of draftsmanship would aid him greatly in other mediums.

Bufford, a sharp-eyed employer, soon recognized the superior workmanship of his newest apprentice and began assigning some of the most exacting productions to him. Soon Homer was turning out more illustrations than the older experienced craftsmen in the shop, where more than 100 worked.

But Homer dearly loved fresh air and the great outdoors and soon came to abhor the acrid odors of the etching chemicals. To break the monotony of the routine at Bufford's, he began getting up at the first light of dawn to go fishing before time to catch the bus into Boston. This practice brought him a stern lecture on punctuality from Mr. Bufford when he missed his bus once and arrived late at the shop. He was never late again.

Homer made friends with two fellow apprentices, Joseph E. Baker and Joseph Foxcroft Cole, and during lunch hours this threesome would visit a small gallery near the lithographic shop. Here, after seeing the work of some of the genre artists of the day, Winslow decided he too would someday paint like that—only better. His dream was formulated and began to take shape.

Luckily, about this time he met a French engraver Bufford had hired for the hurried seasonal holiday business. Damoreau liked young Homer's work and offered to teach him how to draw directly on smooth boxwood blocks—the process used to create the illustrations which appeared in popular pictorial periodicals of the day such as *Ballou's Pictorial Drawing Room Companion* and *Harper's Weekly.*

Homer intuitively realized this training would improve his artistic technique and in time, might free him from Bufford's. In the months following the invitation, Damoreau tutored him evenings in his small sparsely outfitted studio in the Ballou Building. Probably, the most difficult aspect of the whole technique was drawing on the block in reverse. Slowly, under Damoreau's precise instruction, Homer increased his skill. He learned to maintain his own personality and style from original drawing to finished sharp-lined engraving. The association with Damoreau also added to Homer's awareness of the world beyond Boston, for the Frenchman had traveled widely and loved to relate his experiences.

As Homer developed his artistic crafts, he also matured in character and gradually developed a dry wit and quaintness which came through in his drawings. When and where he could, he splashed on a generous serving of humor. These quixotic touches set his works apart from other artists in his field.

On the morning of February 24, 1857, his 21st birthday, Winslow told his parents of his intentions to leave Bufford Lithographic Company. He had come of age that day and his apprenticeship could be ended. His mother was not surprised by his decision; she had long been aware of his discontent. She knew how much he had hated being imprisoned from eight o'clock in the morning until six o'clock at night. His father seemed unalarmed at his son's announcement that he was going to be a free-lance artist. Perhaps the elder Homer's own share of misadventures had taught him to withhold his judgment.

3

Beginning a Career

A determined Winslow Homer set out with Baker and Cole for the Ballou Building and rented a small studio. He had now joined the fraternity of free-lance artists and engravers. His friend Damoreau heartily welcomed him.

Homer's first free-lance drawing appeared about three months later, with the June 6, 1857 issue of *Ballou's.* The simple portrait of one Capt. J. W. Watkins, drawn from a photo was just a beginning but the editors had requested that Homer submit further work.

In the brief time since leaving his bondage at Bufford's, Winslow had stayed busy building a backlog of work. Roaming Boston's bustling downtown avenues, he developed dozens of quickly sketched drawings with a wonderful sense of light and movement. *Ballou's* next printed a whole series of these Boston scenes—they were superior to any previously published and were well accepted.

About this time, a well-known New York publishing firm brought a new periodical to the newsstands. *Harper's Weekly, "A Journal of Civilization,"* delved into extensive coverage of politics and other items of intellectual interest. Baker and Cole eagerly brought Winslow an early issue and encouraged him to pursue *Harper's* as another market for his drawings. After studying the journal carefully, Homer sent a series of four drawings he'd recently made on the campus at Harvard University. The drawings were immediately accepted by *Harper's* editors and were published as a two-page double spread on August 1. Homer was jubilant!

He soon had all the work he could handle providing specialized drawings to fit the particular needs of both *Ballou's* and *Harper's.* The remuneration for this work, however, barely covered his supplies and studio rent.

With the first really cool weather of autumn, Homer decided he needed a new suit. Lacking funds to purchase one in the style he especially wanted, he created a delightful caricature of one of Boston's fashionable and somewhat conceited characters—a Monsieur Paunceloup. With the drawing in hand, he then visited Paunceloup's custom tailor. The latter was delighted with the drawing as an eye-catching window display. A bargain was struck and Homer got his suit of clothes in exchange for the sketch. (This was the first of many such deals Homer made during his 53-year career. In the future, his sketches and watercolors would be used to pay for the services of carpenters and doctors, for vacation lodging and for other incidentals of living. The influence of his father's shrewd bargaining habits were showing.)

When the Homer and Benson

families gathered for the holiday seasons, they richly provided the young artist with many nostalgic settings for his drawings, such as Thanksgiving with Grandfather Benson at Belmont, Mass. Christmas festivities, skating and sleighing scenes around Boston appeared in both *Harper's* and *Ballou's*—with the initials of W. H. tucked neatly into one corner.

Homer began to receive out of town assignments to cover events which the editors of *Ballou's* felt he could handle. Often in these drawings there appeared a well-dressed, slightly built gentleman with a finely shaped mustache. This was Homer's unique manner of "inconspicuously inserting himself into his work," according to Jean Gould in her biography, *Winslow Homer, A Portrait.*[1]

Summer found him sketching at the seashore at Newport, in the woods picknicking and strolling Boston's back bay. *Ballou's* sent him to Cape Cod to cover a series of revival meetings being held there. This bit of reporting appeared in four drawings in the August 21, 1858 issue. In one of these sketches, "slim, short and trim in his city suit" he drew himself, standing outside the revival tent, "calmly sketching the emotional meeting."[2]

While spending Thanksgiving in 1858 at Belmont with the Homer and Benson families, he created a series of four drawings devoted to the holiday preparation and events. These wood engravings appeared two-up on facing pages in *Harper's Weekly* on November 27.[3]

He then received a long list of portrait assignments of prominent citizens and literary dignitaries, usually made from photographs. This was a task Homer thoroughly disliked.

His final assignment for *Ballou's,* a portrait of the Honorable John F. Potter, of Wisconsin, appeared in the December 3, 1859, issue; the publication was being discontinued due to the strong competition from *Harper's.* Since Boston offered no other illustrated newspapers, Homer came up with another of his "sudden unwavering decisions"[4]—he'd move to New York to be near *Harper's.*

1. Jean Gould, *Winslow Homer A Portrait* (New York: Dodd, Mead & Co., 1962), p. 50.

2. Ibid., p. 50. (Author's Note: These wood engravings entitled "Camp Meeting Sketches" appeared in *Ballou's* with this text: ". . . we now pass on to a notice of the engravings on this and the next page, from drawings made expressly for us by Mr. Homer . . .").

3. Author's Note: One of these nostalgic scenes, "Thanksgiving Day—Ways and Means," depicting the arrival of a grocer's wagon and a small flock of turkeys, the plumpest of which is undoubtedly the chosen-bird, was crafted by one of America's leading silversmith's into a handsome etched sterling silver plate by special arrangement with The Metropolitan Museum of Art.

4. Gould, p. 51.

4

On to New York

The young artist found the pre-holiday season in New York an exciting new chapter. Upon arriving in the metropolis, Homer secured room and board with a Mrs. Alexander Cushman at 128 E. 16th Street.

She quickly acquainted him with her assortment of young male boarders. One of whom, Alfred C. Howland, was also a painter who later became an accomplished landscape artist. Howland had a studio nearby on Nassau Street and he assisted Homer the following day in finding a studio in the same area.

Homer soon was busy with assignments from *Harper's Weekly,* whose editors no doubt were delighted that their leading illustrator was now so accessible. Homer depicted New York's teeming streets with their wild horse drawn traffic—emphasizing the pratfalls that lie in wait for the unwary in the city.[1] He managed also to find rustic subject matter in the city's huge new Central Park. "The Sleighing Season—The Upset" and "A Snow Slide in the City" appeared January 14, 1860 in *Harper's;* another drawing, "Skating on the Ladies' Skating-Pond in the Central Park" was in the issue of January 28th.

Harper's wanted Homer on its permanent staff and made him a very generous offer to join them. He chose not to accept it, however. His freedom meant too much to him. (Fifteen years later, George W. Sheldon, an art critic, questioned Homer about this matter of declining *Harper's* seemingly advantageous offer and quoted him thusly: "The slavery at Bufford's was too fresh in my recollection to let me care to bind myself again. From the time I took my nose off that lithographic stone, I have had no master, and never shall have any."[2]

Three blocks from Mrs. Cushman's boardinghouse was the National Academy of Design and Homer enrolled in night classes there. The school, under the tutelage of Professor Thomas Seir Cummings, was reputed to be one of the country's finest but Homer soon found it "had nothing to offer except supervision in copying plaster casts of arms, legs, feet and all the rest."[3]

His determination to be a first-rate artist next led him to Frederick Rondel, who like Damoreau, the Boston engraver, had immigrated from France. From Rondel, Homer learned the rudiments of oil painting such as setting a palette and handling brushes . . . but the instruction was limited, and Homer was largely self-taught. He took perhaps four or five lessons on Saturdays from Rondel at his Dodsworth Building studio.

Homer moved from his Tenth Street studio around the time of his 25th birthday.

It is safe to assume he truly obtained more practical help first-hand, and informally, from some of the many fellow artists who came by his new studio than from Rondel. Among these artists were Eastman Johnson, Roxwell (Russell) Shurtleff and Homer D. Martin. His new studio in the tower of the New York University Building overlooking Washington Parade Ground (now called Washington Square) was accessible via a small, narrow ladder-like staircase. The studio-living quarters offered a magnificent view with full sunlight by daytime and moon and stars at night. His furnishings consisted of a battered wardrobe, cot, stove, "a wobbly three-legged stool," a few chairs and "a model stand he had inherited from the artist who was there before him."[4]

Homer's tower studio became a popular spot for all the artists in the building. Often in the evenings a dozen or more of them and their close friends dropped in to tell stories and drink whatever they could afford.

Homer's brother Charlie came to visit him shortly after the move into the University Building and Charlie also got a taste of the companionship which existed among the group. But the evening's chattering conviviality in the glaring gaslight and smoky studio could not deter Homer from meeting a midnight deadline for *Harper's*—he worked "away in the midst of it all," shouting, "Here, one of you boys, fill my pipe for me! I'm too busy to stop."[5]

It was evident to Charlie that his younger brother was completely satisfied with this bohemian way of life. Charlie was proud of Winslow's artistic success and offered to help him pay for painting lessons but was vehemently refused. Homer wouldn't accept charity—if he couldn't do it his own way, he'd quit the business.

Spring came and with it the War between the States. From his window, Homer could see Union Army recruits drilling on Washington Square's parade grounds. He hoped to remain free from war duty as long as possible but realized he would soon be assigned by *Harper's* to serve as a pictorial reporter at the front.

In the meantime, Homer continued to perfect his skills. He preferred to transfer his drawings onto the wood block himself, and had found that if a staff artist transferred the lines, half of the quality of the original drawing was lost. His style of strong precise drawing suited this line medium—wood block engravings had a distinctive flavor which set them apart. He was gradually achieving more power—advancing in every aspect of his drawings. To alter his course now would have been disastrous to his career.

For years, these early precise black and white Victorian works for *Harper's* and other popular publications were not afforded adequate credit for their aesthetic value. They must be considered adept preparations, leading Homer to his later fine watercolor accomplishments. In a short time he was to become one of America's most imitated illustrators.

When the colorful "Highlanders" of the 79th Regiment of the New York State Militia paraded down the streets of New York City, Homer was on hand to capture the scene. Later he illustrated "The Advance Guard of the Grand Army of the U. S., Crossing the Long Bridge over the Potomac." These were his first two assignments covering the Civil War; and whether he realized the portents of the horrendous battles which were to come, his coverage of the unfolding conflict would supply him the subjects to create some of his most outstanding work and make his name a household word.

Taking a brief respite before his formal war assignment for *Harper's,* Homer locked his studio in mid-June and went to Belmont, Mass. to visit his parents at their new home. He took with him the painter's tools he had purchased—" 'a tin box containing brushes, colors, oils and various equipments,' including a palette and an easel."[6] He was anxious to demonstrate to his mother how Rondel had instructed him to set his palette—with white, yellow ochre, red ochre, permanent blue and raw sienna. Together, they went to the seashore overlooking Massachusetts Bay, near Marshfield, and she posed for two oils, *On the Beach, Marshfield* and *Sand Dune.* Henrietta Homer was proud to be her son's model for his early attempts in oil painting. Her dreams for him appeared to be coming true.

It was now late July and Homer realized as the weeks passed how much he truly missed living in the country and being near his family.

1. Barbara Gelman, *The Wood Engravings of Winslow Homer* (New York: Crown Publishers, Bounty Books, 1969), pp. 38-39.

2. Albert Ten Eyck Gardner, *Winslow Homer American Artist: His World and His Work* (New York: Clarkson N. Potter, Bramhall House, 1961), p. 57.

3. Jean Gould, *Winslow Homer A Portrait* (New York: Dodd, Mead & Co., 1962), p. 54.

4. Ibid., p. 60.

5. William Howe Downes, *The Life and Works of Winslow Homer* (New York and Boston: Houghton Mifflin Co., 1911), p. 35.

6. Gould, p. 61.

5

As a Civil War Artist/Correspondent

His first assignment to the war front came in October, 1861. *Harper's* supplied him with the following letter:

The bearer Mr. Winslow Homer, is a special artist attached to Harper's Weekly, and is at present detailed for duty with the Army of the Potomac. Commanding generals and other persons in authority will confer a favor by granting to Mr. Homer such facilities as the interests of the service will permit for the discharge of his duties as our artist-correspondent.[1]

Pages 808 and 809 in the December 21st issue of *Harper's* carried a double spread illustration depicting soldiers sitting around a campfire while the company's black cook dances merrily, half in the firelight and half in the shadows. Nearly forty figures are included in the sharply-detailed drawing. Homer titled it "A Bivouac Fire on the Potomac." This and other scenes of camp life occupied most of his time. His sketching notebooks became filled with every phase of Army life. These sketches made no effort to obscure the primitiveness of the soldiers' life.

The enlightening experiences he witnessed "were revealing and even sobering."[2] He viewed this great conflict through the eyes of a young man and it had "accelerated under pressure the development of his own powers of observation."[3] Attached to McClellan's Army of the Potomac as artist-correspondent, Homer met the "demands of *Harper's* readers and a complex war."[4] He learned to view, and to portray, life and death objectively.

Christmas 1861 found him trying to sketch with nearly frozen fingers. Camp life was not easy. His brother, Charlie, sent him a copy of M. E. Chevreul's *Laws of the Contrast of Colour and Their Application to the Arts* and he put it to use at once. All his life he used and treasured this book.

The soldiers' arguing and horseplay between battles found them constantly in trouble for infractions of rules, and he drew them being punished. This punishment sometimes consisted of being forced, under guard, to stand atop a wooden barrel holding a sizeable log, rifle-style over their shoulder. Homer's sketch of this episode was titled "The Guard House," and appeared later in a series of 24 souvenir cards, 4 ⅛" × 2 $^{7}/_{16}$", published by L. Prang & Company, Boston.

Prior to his assignment by *Harper's* as an artist-correspondent, Homer had not been in any association with blacks. The company's black cook, Sam, was a jolly fellow. Friendly and good-natured, he would obligingly dance whenever asked to. The wagoners too enjoyed posing at their chores for the artist. Once, "when he came on four of them, snoozing on the ground by the sunny side of the tent, he could not resist getting the picture down on paper."[5] He gave the title *The Bright Side* to the 13″ × 17″ canvas which he ultimately completed of the black teamsters resting in the sun. It appeared in an exhibit in 1865 at the National Academy of Design.

Some of Homer's other fine oils recording his impressions of the Civil War were not completed until the early 1870's. This only proves, however, his strong "determination to finish any project he started; his insistence on tying up loose ends of a period before he left it for another–" One poignant example of the impressions Homer garnered from army life was *A Sketch of a Soldier* (1864). This shows "a mere boy, drafted to fill a man's place in the lines, clad in an ill-fitting man's uniform."[6] Probably the most outstanding success of Homer's war sketches was *Prisoners from the Front* which was exhibited in 1866 at the National Academy. (He'd been elected a member of the Academy in 1865 after exhibiting three paintings.) *Prisoners from the Front* went on to further acclaim when it was exhibited at the Paris International Exposition of 1867.

Homer illustrated the aftermath of the conflict for *Harper's* with such drawings as "Our Watering Places–The Empty Sleeve at Newport."

This wood engraving shows a one-armed veteran being driven along the beach by a woman in a carriage. His drawing, "Thanksgiving Day–Hanging Up the Musket," was used for *Frank Leslie's Illustrated Newspaper* on December 25, 1865.

Rainy Day in Camp,[7] an oil completed in 1871, is considered by one of Homer's biographers as his best war painting. This work depicts the dismal atmosphere of a cold, wet day in an army encampment with five uniformed men standing before a small fire. This and his other oils brought Homer new prominence in America and abroad. This prestige was due largely because of the timely nature of his subject matter but partly because of the individual style of his work.

It was later written that Winslow Homer's Civil War drawings and subsequent oil paintings were created in "simple honesty and with such homely truth as would have given his pictures a historical value quite apart from their artistic merit . . ."[8]

And, again, it came time for another of Winslow's hasty yet unfaltering decisions. He would go to Paris with some of his pals for the opening of the Universal Exposition. He would see how the Europeans responded to his message.

1. Original correspondence in the Homer Collection, Bowdoin College Museum of Art, Brunswick, Maine.

2.-4. Philip C. Beam, *Winslow Homer at Prout's Neck* (Boston and Toronto: Little, Brown and Co., 1966), pp. 7-8.

5. Jean Gould, *Winslow Homer A Portrait* (New York: Dodd, Mead & Co., 1962), p. 67. (Author's Note: The wood engraving of this subject appeared in *Our Young Folks,* July 1866, p. 396, accompanying Thomas Bailey Aldrich's interview with Homer. The famous oil painting by this title is privately owned. *Army Teamsters,* a larger version of *The Bright Side,* is in the permanent collection of the Colby College Art Museum, Waterville, Maine).

6. Beam, p. 7.

7. In the collection of The Metropolitan Museum of Art, New York.

8. William Howe Downes, *The Life and Works of Winslow Homer* (New York and Boston: Houghton Mifflin Co., 1911), pp. 40-49. (Author's Note: Downes considers these works to be "first-hand documents concerning that greatest of our wars, and they illustrate from month to month just how the young artist was acquiring his training").

6

Winslow's Odyssey in France

Winslow and fellow artist, Eugene Benson, decided to hold a joint exhibition and sale of their works to finance their trip abroad and obtained the Leeds and Miner gallery for the last two weeks of November, 1866 for that purpose. Homer showed nineteen paintings. Between the two artists they sold a sufficient number of their works to raise enough money to pay their ship's passage to Europe, "with a fair amount left over for their keep in Paris, if they did not stay too long."[1]

Jean Gould, in her portrait of the artist, states that Homer probably would have been further ahead financially had he not been so fond of clothes. Part of this predilection for sporty attire may be attributed to his having to wear made over or handed down clothes of Charlie's during the leaner years when they were growing up. Among his artist friends, Homer had acquired quite a reputation as being a "natty dresser . . . like a real dandy."[2] Their jokes did not affect him—he truly enjoyed having a sizeable wardrobe and being well-dressed.

The happy group of artists sailed for France early in December. Except for "the taste of 'the salt' "[3] on his Uncle Jim's boat, Homer had never before been on the high seas. The experience was enthralling! He was awed by the thunderous, wild splendor of the wintry, storm-tossed ocean. Warmly scarfed and cloaked, he spent long hours, day and night, out on deck watching every move the crew members made while icy waves crashed over the bow.

Arriving in Paris, Homer promptly located his friend, Albert W. Kelsey, who already had a studio established in Montmartre. Kelsey graciously offered to share it with Homer. Homer fully realized how lucky he was; his money would last longer with such an arrangement—"and he would be right in the midst of the 'bohemian' life."[4]

Paris was an enchanting spot. During the day, they roamed the art galleries and bookstalls watching "sidewalk artists along the Seine" and the young, ambitious art students "earnestly copying the great works"[5] hung in the Louvre. Homer was not concerned about viewing "the Academic works in the official exhibitions" of the Universal Exposition or "even the work of the old masters in the Louvre"[6] . . . he had to see the contemporary painting being done and shown in Paris. (His line drawing entitled "Art Students and Copyists in the Louvre Gallery, Paris," satisfied his interest in that landmark.)[7]

Also living in Paris was another old friend, Joseph Foxcroft Cole, from Boston. At the time Homer left the lithographers in 1859 to go out on his own, Cole left also and came to Paris to study art. For three years Cole had worked in the studio of Lambinet, who was considered one of the "standard gallery gods of the day"[8] in French art circles. Later, Cole made a trip to Italy, visited America, returned eventually to Paris and then had the good luck to fall into the distinguished company of some French painters at the village of Cernay-la-Ville.

Cole next furthered his studies in Emile Jacques' studio in Paris in 1865. When "the ripened results of French art were shown in the great Exposition of 1867," he was on the ground floor. He spoke French fluently, had a "thorough knowledge of the studios and cafes of the Latin Quarter,"[9] knew all the leading French painters of the time, and through him Homer must have gained a very good idea of what was going on in the French art world. Cole undoubtedly made it possible for Winslow to understand the gossip of the studios, and to realize the politics behind the scenes.

As Paris' Spring approached and the weather grew warmer, Homer became restless. "Considered the thing to do,"[10] some of his traveling companions were taking lessons from the local artists. But this was not for him. He wanted to get out of the city and into the country. Equipped with a few recommendations from Cole, he headed alone for Picardy. June, July, and part of August were spent roaming the rich fields and small villages of the French countryside—painting little, but enjoying much. One of his completed landscapes, *Cernay-la-Ville—French Farm,* and a portrait-study, *Girl With Pitchfork,* according to Albert Ten Eyck Gardner, "seem to reflect rather strongly the influence"[11] of France's rebellious avant-garde naturalist artists, especially Manet and Courbet.

At the time of the Exposition, Manet had been the object of adverse critical comment, which strongly attacked his work and subject matter. Yet fifty of his sensational paintings were finally given independent exhibition space outside of the fairgrounds. Cole had probably told Homer about Manet's earlier victimization by the entrenched Jury of the Paris Salon when it had refused three of his submissions. Whereupon, Napoleon III arranged an auxiliary *Salon des Refuses* to be held in conjunction with the regular one, thus allowing the public to decide if the Jury had been wrong or right. That scandal of 1863 marked "the beginning of the modern artist as a man whose creative independence is subject to no check beyond his own conscious."[12]

When Homer got back to Paris near the end of August he thought seriously of returning home, yet really he wanted to stay longer. His work was progressing and he wanted to continue his efforts. Through friends back in New York, he augmented his nearly depleted funds and stayed.

Looking back on Homer's French odyssey, we must realize that there were probably only two aesthetic experiences which left a lasting impression: First, seeing the works of the French naturalists like Manet and Courbet and being asked to analyze their astute handling of color and light; secondly, studying Japanese prints with their "selectivity and sense of design,"[13] which were the rage of Paris.

For at least ten years before going to Paris, Homer had held a theory, which he shared with Courbet, that "If a man wants to be an artist, he must never look at pictures."[14] Courbet had proclaimed, "The Museums should be closed for twenty years, so that today's painters may begin to see the world with their own eyes."[15] Both Manet and Courbet had done their part to open the way for independent artists to express themselves freely in their work. Homer would now pursue the same path.

During his stay in Paris, Homer had returned numerous times to study closely the Japanese art exhibit at the Universal Exposition. He penetrated beneath the surface of these charmingly expressive prints, and found their underlying artistic principles. He discovered in these works a kindred note. Yes, here was what he had been trying to do with his wood engravings! He understood perfectly the message of the Edo designers of wood engravings. But, unlike Whistler or some of his other contemporaries, Homer did not find it necessary to adopt the "obvious decorative externalities of oriental art."[16]

As he left the Continent, Homer was quietly elated that the two paintings he'd submitted for the small American art section of the Exposition were being sent on to Brussels and Antwerp for further exhibitions. He had not received any official recognition in Paris; the only American artist awarded a medal was Frederic E. Church, who was "better known to Europeans than any other American artist of the time."[17]

The *London Art Journal,* however, had given these favorable words: "Certainly most capital for touch, character, and vigor, are a couple of little pictures taken from the recent war, by Mr. Winslow Homer, of New York. The works are real: the artist paints what he has seen and known." And, Paul Mantz, the French art critic, wrote: "Mr. Winslow Homer in justice ought not to be passed by unobserved. There is facial expression and subtility in his 'Prisoners from the Front' . . . we like much also the 'Bright Side' . . . This is a firm precise painting . . ."[18]

These works also received a favorable reception in both Belgium cities. But Homer feared he was gaining the reputation of being simply a genre artist of Civil War scenes—a limitation he had come to abhor.

1.-3. Jean Gould, *Winslow Homer A Portrait* (New York: Dodd, Mead & Co., 1962), p. 88.

4. & 5. Ibid., p. 89.

6. Philip C. Beam, *Winslow Homer at Prout's Neck* (Boston and Toronto: Little, Brown and Co., 1966), p. 9.

7. Appeared in *Harper's Weekly,* January 11, 1868. Reproduced in Barbara Gelman, *The Wood Engravings of Winslow Homer* (New York: Crown Publishers, Bounty Books, 1969), p. 103.

8. & 9. Albert Ten Eyck Gardner, *Winslow Homer American Artist: His World and His Work* (New York: Clarkson N. Potter, Bramhall House, 1961), p. 90.

10. Gould, p. 93.

11. Gardner, p. 90.

12. John Canaday, *Reader's Digest Family Treasury of Great Painters and Great Paintings* (New York: Reader's Digest Assoc., 1965), p. 144.

13. Gould, p. 92.

14. William Howe Downes, *The Life and Works of Winslow Homer* (New York and Boston: Houghton Mifflin Co., 1911), p. 11.

15. Sheldon Cheney, *Reader's Digest Family Treasury of Great Painters and Great Paintings* (New York: Reader's Digest Assoc., 1965), p. 140.

16. Gardner, p. 108.

17. & 18. Downes, p. 57.

7

The Master Artist Evolving

Back in his tower studio in the University Building, Homer completed his two Parisian dancing scenes, transferring them directly to the wood blocks himself. They appeared on facing pages on November 23, 1867, in *Harper's Weekly* with this stern editorial comment: "We shall not venture to look into the abyss on the brink of which these frenzied men and women are dancing, and this too curious crowd of spectators is treading. This is work for the severe and steady eye of the preacher and moralist."[1] (Fun loving Homer was not necessarily in agreement with the editorial comment expressed.)

While on his voyage westward aboard the Cunard Lines' *Catalonia,* Homer had sketched a diagonally-canted on deck scene which he titled "Homeward-Bound." It appeared as a double page in *Harper's Weekly* on December 21, 1867. His biographer, Downes, years later afforded it this interesting description: "The observer is supposed to be looking forward along the port side of the deck, and the ship is rolling considerably to port. One sees no steamer-chairs such as are so universally used nowadays. Several officers are seen on the bridge and one of them is looking through a marine glass at a school of porpoises or dolphins which are disporting themselves off the port bow."[2]

Then, on January 11, 1868, *Harper's Weekly* carried the illustration of art students and copyists in the Louvre. These four works were the only wood engravings to have come out of Winslow's European trip.

Gradually, he found time to add the finishing touches to *Paris Court Yard, The Hayfield* and other of the sixteen oils he'd begun in France. He needed two models, also a cello and violin for *Musical Amateurs* because he wanted to redo the hands which seemed awkward. The figures were "somewhat lacking in modeling." Homer had asked Cole and Baker if they would pose for him, but for one reason or another the three of them never seemed able to get together. This painting, now titled *Amateur Musicians,* is in the Metropolitan Museum's collection. According to J. Nilson Lauvik, an art critic of that period, this work "has the mark of a man who was in dead earnest, even though one might not at that time have been able to predict from this canvas the coming master."[3]

With his previous efforts again brought up-to-date, Homer now "proceeded to work in an entirely new way, and using this new style . . . began to create the unique masterpieces"[4] which would make him famous.

The year 1868 brought Homer many illustration assignments from *Appleton's Journal of Literature, Science and Art, Harper's Bazaar,*

Harper's Weekly, and *Our Young Folks*. He was commissioned by *Galaxy* to create five illustrations for a serial, "Beechdale," written by Marion Harland and running from May through September 1868.

For the National Academy's spring show he exhibited three of his French canvases, *The Studio, Picardie*, and *A Study*. The Academy also insisted on having *Prisoners from the Front* to exhibit—even though it had been shown there only two years before. Its broad acclaim throughout Europe had apparently spurred further interest in America and the public back home had to be satisfied.

Homer's parents happily came from Belmont for the exhibit's opening session. Charlie, just recently married to the former Martha E. French of West Townsend, came also bringing his beautiful bride. Homer instantly adored Mattie—as she was called by everyone in the Homer family. He noted too, and was pleased with, the genuine rapport existing between his mother and her. (She became the daughter Henrietta Homer had probably always wanted.)

Eleven years younger than her new artist brother-in-law, Mattie nonetheless wisely realized the strong bond of affection between the brothers and she never interfered with that relationship. She invited Winslow to come visit them in their new home at Lawrence, Massachusetts. He promised he would as soon as he was able.

A few months later he kept his pledge and on July 25, 1868, *Harper's Weekly* carried a drawing which depicted an army of factory workers, young and old, leaving the Pacific Cotton Mills at Lawrence. The illustration, "New England Factory Life—Bell Time," expressed Homer's sympathy for these textile workers having to be shut up indoors all day long, laboring at monotonous routine toil. Again, his memories of his confinement at Bufford Lithographic were brought back. His thoughts on the subject are obvious from the various woebegone expressions he gave to their faces.

An oil painting, 24 × 38 inches, bearing the title *The Morning Bell* was presumably done about the same time, during his visits with Mattie and Charlie when they were getting settled in their new residence at Lawrence. (A wood engraved version of this title later appeared in the December 13, 1873 issue of *Harper's Weekly.)*

Homer went from Lawrence to Belmont, Mass., visiting with his parents, members of the Benson family and his other cousins. While there, he resumed and completed his studies of the game of croquet which he'd begun several years earlier before his trip to Europe. Another early oil, *Croquet Scene*, (1866) had received favorable comment in *The Nation* for its accuracy in depicting the incredible costumes then being worn by the ladies.

Homer had always had an eye for women, young women especially,[5] and he carefully illustrated them in "a combination of delight and irony." With a keen eye for female fashions, he often painted these young ladies smartly dressed, beribboned and skipping about with "voluminous skirts blowing in the breeze." And, sometimes "on the beach . . . showed a stretch of leg, a clinging of wet cloth to curves, which the critics considered immodest."[6]

About this time, Homer organized a horseback trip into the white mountains with Florence Tryon, one of the cousins in Belmont with whom Homer had always found a close affinity. Florence, an exceptionally pretty young lady, had long been his sounding board; she could get Homer to discuss things which he might otherwise keep strictly to himself. She knew of his great zeal to be a truly fine artist and how impatient he was with this matter. Long ago, she had teased him "for always thinking about business."[7]

Now, in her late 20's and not yet wed, Flossie, as she was nicknamed by her family, understood what Homer was up to when he suggested the trip. Since he'd already asked some of his artist friends in Boston to go, he now wanted her to bring along some of her girl chums to join the entourage. " 'But I suspect," she added, "that your real motive is to make sure you have models.' "[8] Homer admitted that this certainly was the case! (Flossie presumably had been his model for *The Initials,* an oil painting dated 1864.)

To get supplies, Homer rode into Boston with his father. While there he also checked about to see who of the art bunch was around. He found Cole and Baker as well as John Fitch and Homer D. Martin—all were interested in the little expedition he was planning. (From Winslow's subsequent drawings, we can presume that ten persons or five couples were in the party.)

The trip was a tremendous success from his artistic point of view—the mountain light was cool and clear white. Winslow was beginning "to perceive clearly the effects produced by natural light" and, like the French artists he'd recently come in contact with, he was "determined to reproduce open air vision."[9]

During the summer of 1868, Homer also journeyed along the Massachusetts coast between Manchester and Salem. One of his finished efforts of that particular trip—an oil entitled *Manchester Coast*—was shown the following spring at the National Academy of Design's exhibit.

By mid-August, he was back in

his tower studio with an assignment from *Galaxy* for another five serial drawings to run from September through January 1869 as text illustrations for Annie Edwards' story "Susan Fielding." (When this novel was printed later in book form, the frontispiece was simply two of Homer's original blocks reused. It should be made clear in regard to these specific assignments that Homer was, "a rather special sort of illustrator, least happy when adapting his subjects to someone else's text and at his best when he picked his subjects from the simple country life that was familiar and sympathetic to him."[10]

His painting career was advancing steadily. His hand was becoming more deft with the brush. But another turning point, this one in his personal life, was now at hand.

1. Quoted in Jean Gould, *Winslow Homer A Portrait* (New York: Dodd, Mead & Co., 1962), p. 90.

2. William Howe Downes, *The Life and Works of Winslow Homer* (New York and Boston: Houghton Mifflin Co., 1911), p. 60. (Author's Note: The wood engraving "Homeward-Bound" is in the extensive collection of the Cooper-Hewitt Museum, New York).

3. Ibid., p. 61.

4. Albert Ten Eyck Gardner, *Winslow Homer American Artist: His World and His Work* (New York: Clarkson N. Potter, Bramhall House, 1961), p. 90.

5. Gould, p. 106.

6. James Thomas Flexner, *The World of Winslow Homer 1836-1910* (New York: Time Incorporated, 1966), pp. 74, 84-85.

7. Gould, p. 63.

8. Ibid., p. 104.

9. Flexner, p. 75.

10. H.R. Hitchcock's forward to "Winslow Homer Illustrator 1860-1875," *Smith College Museum of Art Exhibition Catalog* (Massachusetts: 1951), p. 6.

8

"Most Interesting Part of My Life . . ."

Homer's biographer, Lloyd Goodrich, touched but briefly on a portion of the artist's life which Homer had kept well hidden from public scrutiny. The facts were revealed in a phone interview to Goodrich by David Rhinelander of the Gloucester Daily *Times.*

In his later years, he was a recluse, a man of few words. But the early accounts show that he was quite a sociable young man until an unhappy love affair turned him away from people. The story is his girl turned him down because he wasn't secure enough, didn't have enough money for her. That hurt him deeply and accounts for his constant worrying about selling his works.[1]

(It must also be considered that financial independence—especially in terms of monetary assistance from his family was ncarly an obsession with Homer.)

The artist himself was probably alluding to his romance when he wrote the following in a letter of 1908 to his earlier biographer, William Howe Downes:

Scarboro, Maine
Augt., 1908.

Dear Sir,

I returned here last Thursday, and I will now answer your letter of June 13th.

It may seem ungrateful to you that after your twenty-five years of hard work in booming my pictures I should not agree with you in regard to that proposed sketch of my life.

But I think that it would probably kill me to have such (a) thing appear, and, as the most interesting part of my life is of no concern to the public, I must decline to give you any particulars in regard to it.[2]

Homer's new female acquaintance was a schoolmistress. They had met initially near the end of August, 1868, at Long Branch, New Jersey, where Homer had gone for a brief holiday with "a bunch of artists" who were camping out along the Atlantic shore. When he arrived, "the season was at its height,"[3] and he immediately set out to capture the resort's carefree gaiety. *Harper's Weekly* would be the recipient periodical of a number of drawings he'd later complete from the oil paintings and sketches he was inspired to do on the scene.

Late one afternoon near sunset, as the tide was beginning to come in, he was strolling along the practically deserted beach when he came upon three women. Evidently, they had been "relaxing on the soft sand, drying off after a dip, when a huge wave rolled toward them and broke before they could get away, drenching them all over again." One of the women, older than the others, had her back to Homer—"watching the wave as it retreated while she waited for the girls."[4]

Homer was being followed by a little stray dog which began barking fiercely at the trio. One girl was bending over from her hips, wringing salt water from her long skirted bathing suit. She had "long blonde hair hanging down limply, covering her face like a mane—probably the cause of the dog's insistent yapping."[5]

Seated to her left on the sand, replacing her bathing shoes was a dark-eyed girl—"her hair completely hidden by one of the tight-fitting caps." The dog's noisy barking was quite obviously upsetting her. Not quite as pretty as his cousin, Florence Tryon, Homer did find "her face was curiously alive, set off by large brown eyes and finely arched brows that not even a frown of annoyance could disfigure."[6]

Swiftly, Homer unfolded his easel from under his arm and hurriedly tried to begin to sketch the scene before him. " 'Oh, be still!' "[7] the seated figure called out to the small, still barking dog. She also looked up at that moment and saw Homer standing there.

" 'Don't move!' he shouted impulsively, which was the wrong command, because then she did move, putting the back of her hand on her hip."[8] (In time, he came to grow quite familiar with this gesture.)

Her frown deepened to a scowl. She spoke with "biting sarcasm" which reminded him somewhat of his former Cambridge school teacher. " 'And just what do you think you're doing, pray?' " He replied assuredly, " 'Sketching—I'm a painter.' "[9] Disbelieving, she and her companion told him that he did not look like one. The oldest member of the threesome had gone on ahead towards the embankment and now beckoned for her companions to follow.

Homer didn't want them to leave. He wanted to make sure he had captured this particular scene as it was just as he came upon it. They couldn't remain to pose and bid him goodbye. His chances of ever again seeing those saucy eyes and that pert face seemed very remote.

The following day he located his artist friends "far down the shore, camping like nomads on a deserted stretch of sand."[10] The convivial group included Homer D. Martin, Eugene Benson, Eastman Johnson, R. M. Shurtleff, and John F. Weir. They urged Homer to stay with them through Saturday for a square dance in the barn owned by the farmer from whom they bought milk and fresh eggs every day or two.

Homer, at first, was not keen on going but changed his mind at the last minute. His mood became jovial when he heard the gay fiddles and he lightheartedly joined the dancers. Midway down the line, he met the brown-eyed girl from the beach. The caller had just given the order to swing partners and Homer grabbed her hand and waist and whirled her in a wide circle.

' "The schoolmistress!' " he called out teasingly. She gasped, ' "How did you know?' "[11]

It was uncanny. Unknowingly, he had guessed her occupation. A shot in the dark which literally had surprised him as well as her. Though she wouldn't give her name or tell him where she was from, he was able to find out that she taught in a country schoolhouse in a small Hudson River village. He didn't dance with her again or see her other than at a distance for the rest of the evening.

Back at the seashore camp, in lamplight, he tried to sketch her face from memory. Try as he did, it wouldn't turn out quite the way he wanted. He must find her again.

Later that fall, he took a brief voyage on a sailing vessel along the New England coast and produced the drawing, "Winter at Sea—Taking in Sail off the Coast." It appeared in *Harper's Weekly* on January 16, 1869.

Homer went to Belmont to spend the Christmas holidays with his family. The old familiar events of the season and outdoor activities provided him with some unique drawings for *Harper's Bazaar*—a more sophisticated offering of Harper and Company. He was much in demand as an illustrator—not only for his incidental genre type drawings—but also for the popular serials published in *Galaxy, Hearth and Home* and other such periodicals. He was not able, therefore, to do as much for *Harper's Weekly* as he had previously. He'd also done a great number of book illustrations for Bryant, Lowell, Tennyson, Whittier, and Lucy Larcom. These brought him greater remuneration than *Harper's* had. And yet, what he truly wanted was to do more painting in oil and watercolor. These paintings were speculative work and he found little time for such. Always there was his studio rent, supplies and travel expenses which had to be paid. The art editors kept him exceedingly busy that winter; this allowed him "little leeway for originality and less time for painting."[12] With the advent of spring however, he had to get out into the country; he had to find the girl from the beach at Long Branch.

In mid-April, Homer left Manhattan and headed north along the Hudson River, with no set itinerary, just going "from hamlet to hamlet—as if drawn by a magnet."[13] As he traveled, he sketched the flocks of chickens being fed by sunbonneted farmers' daughters, cattlc on the hillsides and other idyllic farm settings. He found board where it was available, carrying his equipment with him everywhere he ventured. He kept his expenses to a bare minimum.

In May, Homer was four miles west of Kingston, New York, in the little village of Hurley. The noon air was warm as he came across a knoll and spotted a little red schoolhouse nestled against the green foothills of the Catskills. Children suddenly rushed out of the school's doorway and into the dandelion dotted schoolyard. He waited for the teacher to appear, but she did not. He sat down atop a stone wall to sketch the youngsters in their game of "Snap-the-Whip" and then strolled among the nearby orchard's old gnarled trees to do some other preliminary studies.

Impatiently, he finally approached the schoolhouse and "out of curiosity, he made his way up to the door."[14] He glanced inside. Seated with her back to him was the teacher and across the room, a small boy sat with his face hidden behind a book. Neither was aware of his presence. Homer knocked lightly on the door casing and even before he'd done so, a lighthearted feeling came over him. Somehow he realized that his search was ended.

The teacher now turned. " 'Yes?' " Her expression immediately changed from vexation to complete surprise upon recognizing the figure standing in the doorway. One neat brown eyebrow lifted slightly—" 'Oh' " she gasped. Homer, "with unaffected triumph,"[15] smiled at her broadly. She asked him what he wanted and he replied that he would appreciate it if she would pose for him for awhile, since she already had the unhappy culprit to watch, indicating the boy reading.

Smiling unexpectedly, she said, " 'You're the artist who was at the beach last summer.' "[16] He thought at first that perhaps she would ask him to leave. He relaxed and followed her lead by introducing himself, and showed her a copy of *Harper's Weekly* which contained one of his drawings. She told him she'd not realized the anonymous painter at Long Branch could possibly have been Winslow Homer, the well-known illustrator whose drawings she'd seen on many a newspaper page.

Hesitantly, she agreed to pose for him. His brief sketches that lunch hour later resulted in the painting and drawing entitled *The Noon Recess.* Before the day was over, he'd begun the first of many studies to be made in the next two or three years for the painting, *A Country School.*

When classes were dismissed for

the day, Homer walked her home. Along the way his conversation concerned "his earliest ambitions to become an artist" and how he had once been retained at noontime "for defacing his schoolbooks with drawings."[17] When they reached the farmhouse where she boarded during the school year, she introduced Homer to the farmer's wife. On an impulse, he brashly inquired if there was room for another boarder; upon which he and the school teacher received a measured glance from the matron. Satisfied apparently that their intentions were honorable, she agreed he could stay. Her board and room charges were also considerably lower than at the nearby tavern in the village where Homer had left his gear. This must have been an added reason for him to remain in Hurley for awhile.

Gould writes that this is "the way it had all begun—almost casually, a passing intrigue."[18] Yet when school closed some three weeks later, Homer was still there.

During the previous summer of 1869, Homer had painted an oil, *Artists Sketching in the White Mountains,* which included himself with John Fitch and Homer D. Martin, two of his companions on the trip with Flossie. Now, a year later, he created an illustration from it. In this drawing, however, he deleted his two friends and drew in the schoolmistress, looking over his shoulder watching him work in rapt fascination. The mountain scenery remained much the same since the topography of the two ranges are similar. This latter incident actually took place one early Saturday in June, just prior to their parting for a portion of the summer; she went to visit her parents in Rhinebeck and he returned to his New York studio to complete some works.

Visiting his family at Belmont, Homer made no comment to anyone, not even Charlie or Mattie, of his new heart's interest. He was still unsure of how the plans would work out for the meeting he and the schoolmistress had arranged at Long Branch in August.

They casually met again—as planned—on the boardwalk in front of the hotel where she was staying. His delightful companion had not changed and Homer found her even more congenial than before. So much that he did not hesitate in asking if she would join him and some of his artist friends for an outing. This was the start of their getting to know one another. But before their Long Branch stay was over, they reached "an unspoken understanding."[19]

She proved to be a most workable model—often seeming to enjoy his playful inventions and understanding more rapidly than a professional model what he wanted in the manner of pose and facial expression. Over the period that they were in understanding with one another, the schoolmistress was the main subject for approximately three dozen works, including wood engravings, watercolors and oil paintings done from 1869 through 1877.

In time, she became known to his friends and family as "Win's girl" and it was generally assumed that they would marry one day.

Homer's work at this time also showed evidence of stylistic experimentation. From his initial Long Branch encounter with the schoolmistress and her two friends, he painted an oil on canvas, *High Tide: The Bathers.* (Acquired by the Metropolitan Museum of Art in 1923, a gift from Mrs. William F. Milton, this painting received the critic's appraisal not quite refined when first exhibited.)

Homer went frequently to Hurley during the fall of 1869 and into the winter months of 1870. He convinced himself the place was handy—providing him with the type of subjects he preferred. The scenes he drew were pure, homespun Americana depicting the combination country store-post office as the neighborly gathering spot for farmers and their families to exchange all the latest crop news and gossip; nostalgic harvest scenes; the winter's first heavy snowfall and then, when Spring arrived, the apple trees in bloom.

In his heart, Homer knew he probably "would not have returned to this particular hamlet"[20] so often, if the schoolmistress had not been there.

Alert and charming, "her willful, capricious ways baffled him." She laughed easily, but could just as quickly become "impatient and exasperated." In a flamboyant display of anger "the back of one hand would fly to her hip"—doing so she would be "utterly irresistable" to the artist. Her brown eyes would sparkle! He realized that he was falling under her spell more and more; vaguely he resented his emotions—"he was disturbed about being in love."[21] Perhaps, from the start, he had sensed this was not a transient affair.

During 1870 Homer sent most of his work to *Every Saturday,* a Boston weekly, which for an exclusive on his wood engravings offered a higher fee than *Harper's.* Feeling "he could not afford to turn it down,"[22] he did no work, therefore, for *Harper's* that year. Sadly though, the Boston periodical, like *Ballou's* previously, found staid Boston would not support it and within one year the venture was a failure.

When school closed for the summer, Homer and his schoolmistress joined some of his friends on several excursions. One trip produced the theme for an oil, *Under the Falls, Catskill Mountains,* from which he also later did a wood engraving for *Harper's Weekly* used September 14, 1872. The young woman was "accepted as 'Winslow's friend;' nobody asked any questions, no one voiced any conclusions." Somehow though this very fact of their acceptance troubled him vaguely. He again voiced his deep desires to paint—"to become a painter in the fullest sense"[23] and he told her that eventually he wanted to be financially independent in order to give up illustrating and paint full time. She, in turn, seemed to take it for granted that they would some day be together permanently and hoped they would not have to wait too long.

To Homer it was not just the question of sufficient money which held him back from the responsibilities of becoming a husband; first, he wanted to be master of his art. Desirable as the schoolmistress was, marriage it seemed to her could wait—his painting could not.

He spent part of that summer in the Adirondacks and was again at Long Branch in August. His frequent trips to Hurley were resumed when school opened—he could not stay away. Always he had his sketch pad with him collecting subjects for wood engravings for *Harper's.*

The Metropolitan Museum of Art opened during the winter of 1871 and after viewing the new galleries, Homer found them "no improvement over those in the new Academy Building on 23rd Street."[24] Perhaps he could find a private art dealer to handle his work but until then he continued to exhibit at the Academy. It was there in the spring showing of 1872 that the unvarnished realism of his oil painting, *A Country School,* brought some attention from the critics. When he painted it, he really hadn't wanted the work to designate a particular place—he preferred that it represent to each viewer recollections of his own classrooms of years gone by. Other works included in the Academy's showing were *The Mill, Country Store, Crossing the Pasture,* and *A Rainy Day in Camp.* (With the exception of the last painting most of the scenes were inspirations from his visits to Hurley.) His reputation as a realist and his stature as a painter were increased by these paintings even though there were no immediate buyers.

Later that spring, Homer was notified of a vacancy at the Studio Building, 51 West Tenth Street, New York. Both he and Homer D. Martin had put their names on the waiting list. It was considered an honor to be in the Studios and Winslow was indeed pleased to move into the 15-year-old edifice which had been specifically designed for the working convenience of artists.

Laid out in railcar fashion, the

studios had connecting doors, leading from one studio into another, and above the building's enormous central gallery was a huge domed skylight. Twice yearly the resident-artists would hold a co-operative exhibit. Most of the works were hung in the domed gallery and the connecting studio doors were opened. This afforded a gala open-house atmosphere in which prospective purchasers might wander from studio to studio, getting closely acquainted with the individual artists living and working there. This also meant that within a few hours a select group of hundreds of patrons could view the past six-months' work of these artists. Homer was well aware that these receptions provided him a reliable market for his work and he became as congenial as any of his fellow artists—thus overcoming his usual reticence to avoid contact with the general public and art critics.

During the first two or three years after he moved into the Studios, these preview nights were especially important to Homer. He put a carpet in his studio, bought a brass alcohol urn to brew coffee or tea, borrowed two candelabras and placed them with tall white tapers at opposite ends of the room. His worktable was cleared, allowing room for a fresh floral arrangement of bright blooms (a touch that was noted immediately by Henrietta Homer upon her arrival.) His parents came from Belmont for these events that first year, along with Mattie and Charlie from Townsend. Homer's schoolmistress also came from Hurley to stand beside him, blushing furiously.

On one occasion she'd ordered a special dress for the reception, a lovely mauve colored one with a black-and-white striped petticoat showing beneath its fashionably long, full skirt. Homer placed two camellias with shell-pink petals in her soft brown hair. He was probably very pleased to have her near and see the distinguished array of guests who came to this reception which lasted well into the night.

Things seemed to be going along fairly smoothly for the pair; his parents voiced approval of the girl even though Winslow never told them of any definite plans. Father Homer felt that marriage might improve his son's habits; one which seemed to rankle him most was the artist's habit of breaking the Sabbath by sketching on Sunday.

Living as simply as possible, Homer tried to save money. He found, however, that as soon as he got a bit ahead of the game, there were two or three paintings which needed framing before being offered for exhibition and again his small cash reserve would be gone.

America's five-year-long depression of 1873 set in during the early winter and spring. Suffering "severe 'business reverses' "[25] his parents had to sell their farm in Belmont and move in with some Brooklyn relatives. Charlie was in a new position—chief chemist for Valentine Paint and Varnish Company—and was able to offer his parents some assistance. Homer witnessed his mother's efforts to adjust to cramped quarters with in-laws while standing several hours daily to keep up her watercolors. Together they exhibited that year at the Brooklyn Art Association show, a feat which made them both proud.

Her strong efforts in watercolor prompted Homer to take up the medium in earnest. Some British watercolors shown at the National Academy proved very popular with the American picture buying public and this gave him further incentive. He did one small study of *The Schoolmistress* and was rather pleased with his attempt; he now began to view this medium as an area "to increase his income enough to marry; and in time, if he was successful, he might be able to give up illustrating, which could not satisfy his soul, no matter how skillful he was."[26]

Homer visited with Charlie and Mattie in June and July of 1873; staying with them in a house they'd rented at Gloucester, Mass. The residence was near the ocean and Homer enjoyed centering his attentions on the incidents happening in this fishing port. Going down to the Harbor early each morning, he concentrated his efforts in watercolor, making sketches initially and using a good bit of gouache which caused his first attempts to look mostly like colored drawings. He was pleased that he could work "swiftly and accurately, entirely in the open, recording his impressions spontaneously."[27]

An extensive amount of these impressions, through which he was seemingly recapturing the joys of childhood, became drawings for *Harper's Weekly.* They mainly depicted children along the shoreline, waiting, watching, wading, fishing, building small boats—duplicating the efforts of shipbuilding yards, gathering clams and berry picking, plus the other usual pleasant moments seen on the seashore.

Mattie had written to Homer inviting him to come and stating that the house was spacious enough for the family and a guest—"if Win would like to invite his girl."[28] She agreed to join him in July following her visit with her parents.

When the schoolmistress arrived in July, she came with her arms full of books. It was, therefore, logical for Homer to capture her at her favorite pastime in a number of watercolors—these works are *The New Novel, Reading in the Sun,* and *Woman Reading.* (He also did an oil study of her sitting in a straight chair on the porch which comes close to portraiture.)

Later, one day in mid-July, when the foursome were down at the beach, Homer created a drawing of Mattie and his girl, clad in their ridiculous bathing costumes, as they were about to enter the water. This engraving appeared in *Harper's Weekly* on August 2, 1873, with "The Bathers—From a picture by Winslow Homer" as its caption.[29]

Reference has been frequently made to Homer painting girls "with such great admiration, not only for their beauty, but also for their clothes." Examples of this are seen in one 1877 oil of Mattie, titled *Autumn,* which shows her "walking wistfully alone in the woods;"[30] and in *Girl by the Seacoast* painted in 1888, depicting a summer resident off Prout's Neck, seated on a large vertical rock reading a current novel. The attire of both women is meticulously done with special care given to accentuate detailing of lace neck ruching or scarves, hat trimmings, soutaché braiding and pleated flounces.

Later, back in New York at his studio, he put his drawings on the block and agreed to do some additional illustrating to help his savings. Would he ever have sufficient money put away so that he could feel secure enough to ask the schoolmistress to be his wife? He knew he had many years of struggle ahead to be a truly successful artist whose works would sell readily and for prices they deserved. His technique with watercolor was improving, but he realized he still had a long way to go to attain absolute sureness in this medium.

However, the schoolmistress was impatient with him for not setting the date of their marriage—nor could he quite blame her. Seeing how happy Mattie and Charlie were together had been, perhaps, the main cause for her displeasure with him.

Because of the depression, money was tight, people were just not spending for unnecessary items. His watercolors sold for scant amounts at best—one of his small croquet scenes in oil went for a mere $36. His assignments from *Harper's* became less frequent and by the beginning of 1874, there were many persons jobless in New York. They could be seen in long lines waiting for a free bowl of soup or in the city's jails at night because they had no other place to sleep.

By spring the schoolmistress was without a job—the school in Ulster County having been closed for lack of funds. She did not want to live with her family—she wanted to be married now. Homer listened carefully. He had a fair sum saved despite the bad times—perhaps "they might have a wedding in June."[31]

Just before the showing at the Academy exhibit, Homer decided to place *School Time,* an oil panel, and several other paintings in the spring show. He would ask $200 for it. The American Water Color Society was also running its showing concurrently and Homer's first professional display in that medium brought acclaim from the critics but not one single buyer. He was deeply discouraged, but he would not think of giving up his art career.

Framing of these pictures for the aforementioned showings took most of his savings—he had taken a chance—if he sold one canvas, it would have replaced these funds and given him a profit too. There was a prospective buyer interested in *School Time* who said he could not buy it presently but would later.

June was fast approaching and when the schoolmistress pressed Homer for a definite date, he confessed he had used his savings to frame the pictures for recent exhibitions. " 'How much did they cost?' " she asked. Little realizing the effect his answer might have, he truthfully replied, " 'A hundred and fifty dollars.' "[32]

Upon hearing this sum, for the framing of three pictures, she flew into a rage which provoked a bitter argument between them—ending in bitter tears for her and feelings of pain and guilt for him. He saw her with disillusioned eyes. She hotly told him that he might just as well be wedded to his art, for all that she cared! How could he possibly make her see the full meaning of being an artist!

Impatiently, she broke away from his arms as he was attempting to give an explanation of what it was like to be driven by such a powerful force that even love must wait on its demand. Rushing from his studio, she ran down the stairs and rode off in a rickety old hack she'd hailed.

A month passed and there was little communication between them. She'd given him the "ultimatum of making up his mind by June;"[33] she would wait no longer.

In May, John F. Weir urged him to come up to West Point over the weekend. Homer went for this visit with his fellow artist's family, but his heart was troubled. Could his lovely schoolmistress withstand the endless months of pinched living his mother endured even now? In the predawn hours, he got out of bed and went out into the garden to view the effects of the roseate light of the sun's changing spectrum on the flower beds. Continuing his minute observation, tracing the shade patterns along the walkways, he finally decided that his art was more important to him than anything else in life. If his schoolmistress could not wait for him to provide a proper home for her, he would never marry.

What followed between Homer and his girl was not as difficult as he might have surmised. Something had happened between them—there would be no compromise. Neither had any intention of giving in to the other. They would simply go their own separate ways. Before they parted, though, Homer requested one last quiet day in the country and asked her to wear her mauve dress with its black-and-white striped petticoat. She protested saying that it was just a tattered old thing. He insisted and she complied to this one final request.

The setting was serene and her mood was whimsical. From a deck of cards in her purse, she proceeded to tell his fortune out of pure mischief. Homer had seated her on a slight hill which offered a green backdrop of shrubs and began to set his palette, according to Gould. The schoolmistress accounted his future:

" 'You will live to be an old man—a famous, lonely, crusty old man. You will have houses galore and no children to live in them. Your only companions will be your paintbrush and your pipe.' "[34]

Her eyes were flashing and the familiar gesture of one hand on her hip pressing hard against the side, palm up, came almost instantly. Homer listened, smiled and asked her to hold the cards a trifle higher.

It was a long time before that particular portrait was completed. Finally, at the Academy's spring show in 1876 it appeared with some of his other works. It was barely noticed and drew comments only for its rather provocative title, *Shall I Tell Your Fortune?*—a study that he retained in his own possession 'til the day he died.[35]

1. David Rhinelander, "Winslow Homer Loved Cape Ann" *Gloucester Daily Times* (Gloucester, Massachusetts) December 12, 1962.

2. William Howe Downes, *The Life and Works of Winslow Homer* (New York and Boston: Houghton Mifflin Co., 1911), pp. 224-225.

3-16. Jean Gould, *Winslow Homer A Portrait* (New York: Dodd, Mead & Co., 1962), pp. 114-124. (See also— Barbara Gelman, *The Wood Engravings of Winslow Homer* (New York: Crown Publishers, Bounty Books, 1969), p. 146. Reproduction of the wood engraving "High Tide," which later appeared in *Every Saturday,* August 6, 1870, p. 504, is from Homer's initial encounter with the schoolmistress at Long Branch).

17. Gould, p. 124. (See also—James Thomas Flexner, *The World of Winslow Homer 1836-1910)* (New York: Time Incoporated, 1966), p. 30.

18-24. Gould, pp. 125-134.

25-28. Ibid., pp. 141-142.

29. Gelman, p. 165.

30. The oil painting, *Autumn,* is in the collection of Mr. & Mrs. Paul Mellon.

31.-34. Gould, pp. 148-152.

35. Author's Note: Although romantic rumors have persisted about Homer for many years, none of his previous biographers have documented the identity of the schoolmistress. Until positive proof has determined it, her name will therefore be kept unpublicized.

9

Getting Away from the Past: Petersburg, Virginia:

Homer knew his affair of the heart was over and that he needed to have a change of scene to restore his own inner perspectives. His life would go on and would assume the pattern he chose for it. Where might he go right now, though, to get away from all the memories which were so close (even here in this very studio)? His funds were too depleted to allow him the luxury of returning to Europe, so that was out!

While reassembling some of his drawings and character studies, he came across some of his Civil War sketches of Negroes drawn near Petersburg, Virginia, showing various subjects picking cotton in the few remaining unburned fields. He wondered what it would be like down there now, nearly ten years after the war. It would be interesting to find out, he decided.

After his 39th birthday in February, IIomer packcd his art cquipmcnt and grip of clothes and boarded a train. (Before leaving New York, he also apparently went by *Harper's Weekly* and gave notice to its editors that there would be no more line illustrations bearing his signature forthcoming—no more would he be harassed by engraving deadlines!)

His travel plans were perhaps not definite. But in the back of his mind, he knew he'd find what he required in the manner of a challenge and change of pace.

Arriving at Petersburg, Winslow rented a room in a small hotel on the town's main thoroughfare. Soon after unpacking he located the black community and sought to make friends with "these people over whom the war had been fought."[1] He soon found that life was filled with great hardships for the blacks—the 14th Ammendment had brought them freedom but that had meant only their liberation from a particular master. Economically, they were now forced to shift for themselves—to earn sufficient amounts each day to keep their bodies alive.

Despite the problems they faced, Homer noticed their spirits were high—their faces expressed patience which could outlast prejudice. One wizened, old black called Uncle Ned had a small patch of land and he worked the soil with pride. He was eager to make each small gain a milestone toward equality. As soon as Homer could get his equipment set up, he did a painting and entitled it *Uncle Ned at Home.*

Each morning, Homer took breakfast in the hotel's dining room and then went down to paint among these simple, friendly dark-skinned people who were happy to pose for him. Some of the town's white, socially prominent female residents, upon hearing there was an artist in their midst, sent invitations to Homer at the hotel. Since he was not often present, he could not accept them personally; written replies did the trick.

One afternoon, after leaving Uncle Ned's cabin, Homer went down to another cabin to complete a subject of a mother with her two daughters. Just as he arrived, he looked up to see an older white woman whom he recognized to be one of the town's social leaders. She was, quite obviously, the previous owner of these former slaves. Homer requested that she pose with them, "just as she had been standing there, conversing with them in prim, threadbare elegance."[2] His arrival and subsequent request caused her surprise and discomfort but she could not refuse. In a few minutes he was able to secure a sufficient impression to paint one of his most successful pictures, *A Visit from the Old Mistress*, now in the collection of the National Gallery of Fine Arts, Washington, D.C.

Homer returned to his New York studio and there completed three oils: *A Happy Family in Virginia, The Unruly Calf* and *Weaning the Calf.* (Another watercolor subject of this visit, *Springtime in Virginia*, appeared later in E. P. Richardson's *A Short History of Painting in America.)*

Homer's younger brother, Arthur, was to be married in July to Alice Patch and following the wedding the young couple let the artist know he was welcome to join them at Prout's Neck, Maine, where they had made reservations at Checkley House, then the area's only available accommodation. He enjoyed getting acquainted with his new sister-in-law and did a painting of them together which he presented as a wedding gift. He titled the work *The Honeymoon* and signed and dated it.

Later that fall season, he exhibited *The Course of True Love, Landscape Milking Time* and *Uncle Ned at Home* at the National Academy of Design.

The artist returned to Petersburg in 1876 to paint *The Watermelon Boys*, get material for *The Carnival* (1877), *Sunday Morning in Virginia* (1877), and *The Pumpkin Patch* (1878). This series of oils and watercolors made Winslow among the most individualistic of painters.

Never before had the American Negro been portrayed with such sympathy and understanding. He depicted the "humor in their actions without making fun of them;" pointing out the miserable conditions in which they lived "without overpitying them;" revealing their strongly religious, poetic nature "without glorifying them." With his always objective eye, Homer "saw much artistic value in a race that had heretofore been represented principally in caricature."[3] He changed the concept.

Fellow artists Alden Weir and John LaFarge gave him great encouragement in this latest work—they recognized his "distinctive genius which, though still in the budding stage, gave promise of great flowering."[4] Homer D. Martin added his support and Winslow deeply appreciated the heartening comments from all three of his colleagues. Homer found little time or inclination to feel sorry for himself. His close-knit family relationship with his two brothers and their wives, his parents and his fast friendships with co-residents at the Tenth Street Studios kept him from being lonely. After all, this was the path he had elected to follow. He would not indulge self-pity or recrimination.

1. & 2. Jean Gould, *Winslow Homer A Portrait* (New York: Dodd, Mead & Co., 1962), pp. 155-156.

3. Ibid., p. 160. [See also—James Thomas Flexner, *The World of Winslow Homer 1836-1910)* (New York: Time Incorporated, 1966), pp. 86-87, and John Wilmerding, *Winslow Homer* (New York, Washington, and London: Praeger Publishers, 1972), p. 94].

4. Gould, p. 161.

10

Summer Daze of 1878: Houghton Farm

Part of Homer's summer of 1878 was spent near Cornwall, New York, at Houghton Farm, Mountainville, as a guest of Lawson Valentine, Charles S. Homer, Jr.'s business partner. The two Valentine brothers, Lawson and Henry, understood Homer's aversion to the public and its prying; they were going to be gone for the summer, so Winslow jumped at this opportunity to have complete freedom in this pastoral retreat.

Here, at Houghton Farm, according to Donelson F. Hoopes, the artist was caught firmly "in the spell of nostalgia for the eighteenth century, a phenomenon which swept the county following the Centennial Exposition of 1876."[1] Homer painted over two dozen watercolors with his preferred subject being a little shepherdess, her flock, the surrounding hillside, stiles, fences and various other pastorally-idyllic settings.

It is fascinating how this shepherdess subject evolved: Shortly after his arrival at Houghton Farm, Homer discovered that the little slip of a shepherdess in her early teens who tended the flock in a lackadaisical manner was the daughter of a poor mountaineer nearby. Winslow made arrangements with the father for her to pose for his sketching while watching the sheep. This invitation really thrilled the half-grown young lady and when she arrived for her first sitting, she came attired in her Sunday-go-to-meeting clothes, all starched and beruffled. The artist was not pleased and sent her home to change into her usual smock and sunbonnet. She did so—bewildered and crestfallen. Winslow, realizing the effect his brusqueness had had on her, was quite sorry.

So, artist and model made a bargain; if she would pose for him in her loose fitting calico and limp sunbonnet, he would get her a proper shepherdess costume and paint her in it later on.

He sent away for a shepherd's crook, buckled shoes, a straw hat with streaming blue ribbons and a dress with laced bodice and flounces. This whole thing pleased her enormously and she became the perfect model in her 18th century outfit.

The resulting charcoal studies and spontaneous on the spot watercolors captivated the public and critics alike. Homer was surprised, yet pleased. He'd not expected such acclaim—he did not feel these works merited the amount of favorable comments they evoked. His favorite little shepherdess of Cornwall gained him recognition, in America as well as in Europe, for the naturalistic quality of his work.

Many of the Houghton Farm subjects were promptly purchased by the Valentines—thus assisting Homer over the financial battle he had been fighting since he had ceased making wood engravings for *Harper's* and the other periodicals. (Most of these shepherdess works, along with some other earlier subjects from around Hurley, New York, are now in the permanent collection at the Colby College Art Museum, Waterville, Maine—the contribution of the Harold T. Pulsifer Estate.

Later that autumn—anxious not to acquire a name for himself as "a painter of shepherdesses," Winslow talked Charlie into accompanying him on a hunting-fishing trip to Keene Valley. On the very first night of their trip, as they camped "beside the inscrutably black forest" he painted what turned out to be probably his most important oil of that year. *Campfire,* in unconventional simplicity depicts two male figures in the faint light of their campfire, surrounded by forest darkness. Sparks fly up in daringly novel patterns while nearby weeds lean in toward the flames in many delicate patterns. One almost hears the blaze crackling against the nocturnal silence. Gardner termed this painting "among Homer's best works, and, being there, it must also then stand among the very best American painting of the time."[2]

1. Donelson F. Hoopes, *Winslow Homer Watercolors* (New York: Watson-Guptill Publications, 1969; third printing 1971), p. 22.

2. Albert Ten Eyck Gardner, *Winslow Homer American Artist: His World and His Work* (New York: Clarkson N. Potter, Bramhall House, 1961), p. 204.

11

Second Gloucester Visit: Summer 1880

Homer was continually being lured into the world of watercolor. During the early spring of 1880, with sufficient funds salted away from recent sales, he became more than ever determined to carry through with a plan he'd been unable to proceed with the previous summer, due to an unfortunate accident.

It seems that while visiting with Mattie and Charlie at their West Townsend home, Homer broke his arm. This unforeseen setback delayed his schedule for further developing his watercolor technique. It happened in June, just as the three were preparing for their respective vacation trips—Homer to go his way and Mattie and Charlie to go theirs. (There is no record of how the break actually happened or to which arm.)

The family physician in Townsend, Dr. Royal B. Boynton, came over and set the bone, which he cautioned Homer would require about six weeks to heal. Bitterly disappointed with this delay, Homer allowed his feelings to become quite apparent. He did not, however, want Mattie and Charlie to change their vacation plans. Dr. Boynton graciously offered the artist a place to stay in his home and Homer could do nothing but accept with gratitude.

The doctor's young son, Henry, soon became Homer's delighted and devoted fishing companion. Daily, at 11 a.m. and 3 p.m., this twosome would take to the back waters of the Squannacook River and fish for about an hour. Each time Winslow would adroitly cast into the lily pads and, in spite of his broken arm, catch his fair share of pickerel. Henry grew to have great respect for "Mr. Homer's" fishing skill!

As the bone began knitting, Homer took up his sketch pad again and later his paintbrushes. Before he left Townsend, three paintings were nearly completed. One of these, *Woman Driving Geese,* Homer later inscribed and sent to Dr. Boynton as a token of his gratitude for the care and kindness he had received in the physician's home.

Now, a year later, he was well and completely ready to further his growing mastery of his profession. With the aid of his good friend Samuel Preston, Homer arranged lodging for himself for May through September with the lighthouse keeper Merrill and his wife on Ten Pound Island in Gloucester harbor. There, he again spent the summer months—a period without any intrusions. He rowed over to the village only when absolutely necessary for supplies.

Downes, in his 1911 biography, described this as a fruitful time for the artist:

The freedom from intrusion

which he found in this little spot was precisely to his liking, and here he painted a large number of watercolors of uniform size but of a wide range of boldly conceived and vigorously executed subjects.[1]

According to Gardner, one might get the impression a greater part of Homer's life seems "to have been a vacation and his work as an artist—in watercolor at any rate—was a congenial form of vacation pastime."[2]

Considered objectively, Homer's selected subject matter reflects not only his own interests (boating and fishing), but also that same trend in American life which began to grow following the Civil War and has continued to flourish ever since.

Fortunately, the three Homer brothers were financially situated so that they could spend their summers or winters where they chose: Gloucester, Mass., deep in the Adirondack Mountains, or eventually at a wilderness camp of Quebec province. Wherever and whenever Homer went, both his watercolor box and his fishing tackle went also. He had a knack for being in the proper locations with people, especially the reliable guides, who respected his quiet nature and were—like himself—men of few words.

One of the people with whom Homer became acquainted during the summer of 1880 was George J. Marsh who lived at 4 Leonard Street and shared with Homer a love of the sea. Marsh owned a small sloop and Homer painted a watercolor of him sailing it in Annisquam Harbor. (This subject is now in the collection of the Indianapolis Museum of Art.)

Working swiftly, from sunrise 'til sunset, Winslow made his watercolor sketchings standing on the island's shores studying the effects of light. Some of the titles of his watercolors coming out of this summer's stay are: *Children Playing Under Gloucester Wharf; Gloucester Harbor, Fishing Fleet; Gloucester Schooner and Sloop; The Green Dory; Boys Beaching a Boat; The Sloop Kulinda; Moonlight Schooners, Gloucester; Sunset at Gloucester,* dated (8/25/'80) and *Wreck Near Gloucester.*

Since Gloucester is even today a favorite spot for vacationers, we can safely assume that Homer was joined by his family for a short time during his sojourn there. Two young women appear in a number of Homer's works of that summer—probably Mattie and Alice Homer.

Homer had by now, according to Wilmerding, "fully mastered the fluid and improvisory nature of the medium (watercolor) and worked quickly out-of-doors in the bright sunlight."[3]

This recent biographer continues by pointing out: "The application of his washes is broader and more confident than in his previous watercolors. The less cluttered designs and more detached viewpoints of these pictures reflect Homer's increasing thoughtfulness about his art. Gradually disappearing are the concentrated narrative or anecdotal aspects of his earlier work, to be replaced by a greater interest in formal matters, on the one hand, and more serious, even philosophical, subject matter, on the other."[4]

Bostonians greatly admired his Gloucester works—this was familiar territory to them. New Yorkers, on the other hand, had hailed the Houghton Farm pictures, but were decidedly cooler in appraising these. The Gloucester works appeared in an exhibition at Doll & Richards' galleries in December and before the show closed "at least a third of the watercolors were sold—at moderate prices, starting at $50—for a return of $1400 after commissions." Homer's labors of that summer were, therefore, "of practical as well as artistic value."[5]

For the American Water Color Society the following spring, he sent twenty or more for exhibition—"only to have them slighted, if not ignored."[6] They were hung poorly—in corners, over doorways and on the skyline. This produced very lukewarm reception from the critics with very few good impressions and most uncomplimentary write-ups. Only George W. Sheldon, as in his previous "Sketches and Studies," came to the artist's rescue and in a second interview allowed Homer to reaffirm his own convictions; firmly advocating the need to paint out of doors:

I prefer every time a picture composed and painted outdoors. The thing is done without your knowing it. Very much of the work now done in studios should be done in the open air. This making studies and then taking them home to use them is only half right. You get composition, but you lose freshness; you miss the subtle and, to the artist, the finer characteristics of the scene itself—it is impossible to paint an outdoor figure in studio light with any degree of certainty. Outdoors you have the sky overhead giving one light, then the reflected light from whatever reflects, then the direct light of the sun, so that, in the blending and suffusing of these several luminations, there is no such thing as a line to be seen anywhere. I can tell in a second if an outdoor picture with figures has been painted in a studio. When there is any sunlight in it, the shadows are not sharp enough; and, when it is an overcast day, the shadows are too positive. Yet you see these faults constantly in pictures in the exhibitions, and you know that they are bad. Nor can they be avoided when such work is done indoors. By the nature of the case the light in the studio must be emphasized at some point or part of the figure; the very fact that there are walls around the painter which shut out the sky shows this. I couldn't even copy in a studio a picture made outdoors; I shouldn't know where the colors came from, nor how to reproduce them. I couldn't possibly do it. Yet an attempted copy might be more satisfactory to the public, because (it is) more like a made picture.[7]

This was probably the lengthiest, most involved statement he would ever make concerning his truly intense feeling for nature, "his unceasing quest for truth in reproducing the magic and wonder of outdoor light in art."[8] Literally, his soul had been bared!

1. William Howe Downes, *The Life and Works of Winslow Homer* (New York and Boston: Houghton Mifflin Co., 1911) p. 75.

2. Albert Ten Eyck Gardner, *Winslow Homer American Artist: His World and His Work* (New York: Clarkson N. Potter, Bramhall House, 1961), p. 162.

3. & 4. John Wilmerding, *Winslow Homer* (New York, Washington, and London: Praeger Publishers, 1972), pp. 90-95.

5.-8. Jean Gould, *Winslow Homer A Portrait* (New York: Dodd, Mead & Co., 1962), pp. 186-189.

12

On to Tynemouth: 1881-1882

Homer, according to Kenyon Cox in a 1914 essay, was at the crossroads in his career, prior to his trip to England in the spring of 1881.

Beam later assessed the situation in this manner:

By continuing the conventions that his public had come to accept, he could ensure himself a steady income and an established position. The alternative would be to attempt to advance his art by experimenting with new forms and conceptions . . . seeking new challenges on ever higher levels. To continue to paint as he had for the past ten years would have meant popularity and security, but probably an end to growth. To venture into untried fields was to incur the risk of misunderstanding, neglect and failure. With the new master, as with the old, it was a period of transition and trial that proved decisive.[1]

John Wilmerding, one of Homer's more recent biographers, states the facts with utter simplicity:

"By the end of the 1870's, Homer had reached a watershed in his life and his art."[2]

Apparently, A.T. Gardner felt that Homer's story lacked dramatic narrative and was without pace or tension.[3] To those who have scrutinized certain elements of his personal life and his struggle not to be influenced by sharp-tongued critics, nothing of course, could be farther from reality. Homer's story is one of continual tension from various sources—sources which he continually fought.

A Doll & Richards sale of Homer's Gloucester works in December, 1880, proved successful; putting sufficient funds back into his savings for him to look again for new horizons. He waited until a spring exhibit at the American Water Color Society, which brought out 23 of his recent works, had closed. He then took his savings, gathered up his gear and left for England.

After a lengthy crossing, Homer arrived finally at Cullercoats, near Tynemouth. He had learned about the village from his fishing friends at Gloucester Bay the previous summer. He found a little cottage surrounded by a formidable wall with a gate to lock, assuring complete privacy. He had made up his mind to live by himself. Homer's art was becoming his religion; it might perhaps also be said that, like the Spanish artist El Greco, he would eventually come to live "much by himself and within himself."[4]

Surprisingly, when Homer set out to paint at Tynemouth, he found that he was not especially interested in painting the gaiety and colorful life of this summer resort which the Tyne River as a center for yacht racing had become. Other artists, especially thc popular British ones like J.M.W. Turner and John Constable, had painted in this coastal area. But they depicted the gay carefree vacation life it offered. Homer saw this setting differently. He viewed it soberly and painted its pervasive grayness and monochromy.

He chose a broader scope for his Tynemouth subject matter—the powerful effects of the North Sea, the magnitude of clouds, fog and mist, and the sturdy fisherwomen of the area. His figures became mature—no longer the youthful maidens of Houghton Farm or Hurley's barnyard girls. They were now isolated—no more the adolescent little boy groupings which he had painted at Gloucester Harbor. The stance of his figures was more sculpturesque—their step slowed. Brooding atmospheric effects with wave dashed rocks became more evident.

Homer witnessed firsthand and appreciated deeply the perils these people dealt with daily. Initially, with charcoal, pen and ink, he captured on paper the drama of the North Sea's effect on Tynemouth's native populace. Later, he transferred his attention to his watercolor box and translated the rescue scenes which were frequent occurrences. New England's chronicler of homespun country life found at Tynemouth a far more heroic concept of humanity than he'd ever before experienced. And it came through, shining clear, in his works. "His style developed a new largeness and power: forms were rounder, color more subtle and functional, compositions more deliberate."[5]

When these Tynemouth watercolor paintings were exhibited in February, 1883, at the American Water Color Society, they were the sensation of the show, bringing Homer the first general acceptance by critics and public alike.

When he eventually settled down at Scarborough, Maine, later in 1883, Winslow was fortunate enough to find a number of young women who had the bearing and strong resemblance to some of the figures he'd used at Cullercoats. Cora Googins Sanborn, Sadie Sylvester and Ida Meserve Harding were his principle models during the mid to late 1880's when he presumably completed his Tynemouth subjects.

Homer found companionship during his lengthy stay in England by purchasing a small black and white terrier puppy which he named Sam. They became, in the years ahead, nearly inseparable.

Returning to his New York studio late in November, Homer continued painting a monumental oil he'd begun in England. This large canvas, titled *The Coming Away of the Gale,* depicts a single fisherwoman with her baby tied onto her back in American-Indian papoose fashion. She is striding against strong winds toward a life brigade house, where volunteers are preparing to launch a lifeboat into the heaving seas. Completed and exhibited two months later in the National Academy's spring show, this work was prominently hung and became an attraction of the exhibit. Unfortunately, every reviewer who viewed it found it offensive. They commented on its "theatrical arrangement of the figures" and termed it a "disappointment."[6]

While appearing to shrug off these sharp criticisms with his customary indifference to the press, Homer was actually deeply hurt and bitterly disappointed. The canvas was put away and never again exhibited. Homer, the artist, hid his rejected painting in the same manner as Homer, the man, hid his private feelings from public view.

He might have destroyed the canvas, but perhaps it was his Yankee thriftiness that prevented him from doing so. (Ten years later, the artist took the work out of hiding, repainted the areas of the lifeboat launching and life brigade house and left only the central figure of the fisherwoman and her baby. With a minor livening up of the color scheme, he then retitled the painting *A Great Gale.* It sold quickly to Thomas B. Clarke, who sent it to Chicago for exhibition at the 1893 World's Fair and drew high praise and admiration as a new work. It also brought Homer a coveted gold medal!

1. Philip C. Beam, *Winslow Homer at Prout's Neck* (Boston and Toronto: Little, Brown and Co., 1966), p. 15.

2. John Wilmerding, *Winslow Homer* (New York, Washington, and London: Praeger Publishers, 1972), p. 96. [See also—Donelson F. Hoopes, *Winslow Homer Watercolors* (New York: Watson-Guptill Publications, 1969; Third Printing 1971), p. 17].

3. Albert Ten Eyck Gardner, *Winslow Homer American Artist: His World and His Work* (New York: Clarkson N. Potter, Bramhall House, 1961), p. 60.

4. George Kent, *Reader's Digest Family Treasury of Great Painters and Great Paintings* (New York: Reader's Digest Assoc., 1965), p. 84.

5. Lloyd Goodrich, "Pictorial Poet of the Sea and Forest," *Perspectives U.S.A. #14* (Winter 1956), p. 48.

6. Lloyd Goodrich, "A 'Lost' Winslow Homer," Worcester Art Museum *Annual,* vol. 3, (Massachusetts: 1937-1938), p. 70.

13

Becoming a Maine Artist

The year 1883 was another restless one for Homer. His two-year sojourn at Tynemouth was inspiring him to make further studies of the region's fisherfolk. His studio at 80 East Washington Square in New York City did not, however, seem the logical location in which to complete the scenes his vision harbored. Nor could he find models who sufficiently resembled the grace and bearing of the stalwart wives of Tynemouth. His storehouse of sketches begun in England had to be completed.

Prout's Neck, Maine, seemed to be the answer. Prout's Neck was a place of remarkable beauty where Arthur B. Homer, the youngest of the brothers, took his bride, Alice Patch, for their honeymoon in 1875. Winslow Homer visited during that summer at the Willows Hotel and discovered the spectacular views the Neck afforded when an easterly storm broke against the rocky cliffs, sending the surf fuming up furiously. Here, 109-miles from Boston and 12-miles south of Portland, were the kind of rocks and seas for which Homer had longed.

The entire Homer family fell in love with Prout's Neck. They began acquiring property and constructing homes there in 1882 with Arthur building a cottage, which he officially named *El Rancho* to commemorate his many years of business in Galveston, Texas.

Charles S. Homer, Sr. followed suit. Driving a shrewd bargain, he acquired a large parcel of land on which a stable and the frame of what was to have been a large boardinghouse were built. When the dwelling was completed in 1883, he christened it the *Ark*.

Although a painting room had been provided for Homer at the big house, he chose instead the stable. He promptly hired an architect, had the building moved closer to the cliffs and then specifically remodeled to suit his needs and desires. This was his retreat—near enough to the family he loved—but still affording him the complete privacy his work required.

Solitude had become an absolute necessity to him. He had found it in Tynemouth and on Ten Pound Island in Gloucester Harbor. Now, at Prout's Neck, he would settle in a place of his own and find solitude. Because he preferred solitude to the bustling crowds of people in the city, many people developed the opinion that Homer was strictly a loner. He enjoyed people in their proper time and place. And, as an extremely resourceful man, he appreciated being left alone to work out his painting projects. When, and if, his family did interfere with his schedule, he would retreat to his studio.

Many of Homer's most famous

The artist's water color box of Winson & Newton cake colors with his name and address. Courtesy of the Homer Collection, Bowdoin College Museum of Art, Brunswick, Maine.

Maine watercolors were painted outdoors in the vicinity of his converted stable-studio at Prout's Neck. One favorite spot overlooked Ferry Beach on Saco Bay. Here he had a small shack on a slight rise; he used this for nearly 20 years as a storage place for his painting materials. It also provided refuge in case of a sudden squall. After being broken into a number of times during winter, it was finally sold to a neighbor.

After his two-year absence in England, Winslow Homer welcomed the chance to be near his family again—especially his mother whom he adored. Throughout the years of his early apprenticeship and later free lancing illustrator days, Henrietta had continually demonstrated deep understanding and strong faith, not only in his talent, but also in his modus operandi.

Their mutual admiration for each other's work was ever evident. Over the years she retained his letters and later passed them on to Mattie, Charles Jr.'s wife, for safekeeping. Winslow kept a group of her floral watercolors—these being the only works of another artist that he ever maintained.

He visited Henrietta's bedside when she was ill in 1883 and he wrote to Mattie of her condition:

I find Mother very much better.

In fact to me she seems very well.[1]

A further portion of his letter expresses his 74-year-old mother's attitude, concerning the need for independence, which in later years proved to be nearly identical to his.

She wishes it understood that when she gets old her greatest comfort will be to do as she likes—[2]

Her death on April 27, 1884, was a great loss to Homer. He had enjoyed her sense of humor and had welcomed her appreciation for his work.

With his family, Winslow was a deeply affectionate, but undemonstrative person. His father, brothers and sisters-in-law (especially Mattie) were vital to his life. His letters to them may have been brief, but they were frequent and continually filled with expressions of genuine appreciation for small favors and gifts sent to him throughout the years. This exchange was a two-way street and he often sent them items for no apparent occasion or reason other than that he felt like doing it. They did the same. Winslow was a decidely thoughtful man, he seemingly never failed to say or write his own witty bread-and-butter bits of correspondence.

By late summer 1884, Homer was settled at Prout's Neck and, as the tourists and summer residents began to depart, around the first of September, he grew to love his new home even more. There was, however, a fly-in-the-ointment. It came in the form of a decision made in Cumberland County's legal circle that the newcomer artist on Saco Bay might be a good juror.

This particular civic duty was something Thoreau-like Winslow Homer abhored. He had his fill of court procedures while on assignment for *Harper's Weekly* and while serving on a jury in Belmont, Mass. His distaste was expressed in a courtroom illustration depicting jurors listening carefully to counsel during a Supreme Court case. The editor praised the remarkable "delineation of character" and "faithful representation of life in the arena of jurisprudence."[3] To Winslow, this was one arena he wished to remain as far away from as possible. He would not be harassed so he packed his bags, put Sam, his English terrier, to board at West Point House down the road from his studio, went down to Boston by train to confer with his agent and oversaw the installation of an exhibition of his black-and-white studies. He also saw to it that his father, Charles S. Homer, Sr., was properly situated at the American House, a favorite hotel of the family. Next, he booked passage by steamship to Nassau on New Providence Island in the Bahamas.

1. & 2. Quoted in Philip C. Beam, *Winslow Homer at Prout's Neck* (Boston and Toronto: Little, Brown and Co., 1966), p. 36. (Author's Note: Original correspondence in the Homer Collection, Bowdoin College Museum of Art, Brunswick, Maine).

3. "Jurors Listening to Counsel," *Harper's Weekly*, (February 20, 1869), p. 120. [See also—Barbara Gelman, *The Wood Engravings of Winslow Homer* (New York: Crown Publishers, Bounty Books, 1969), p. 119].

14

Discovering the Tropics

Each morning in the Bahamas brought new discoveries for Homer. A rainbow riot of exotic tropical flowers greeted him immediately outside his window. Pure blue skies and bright sunlight deepened the squint-wrinkles at the corners of his hazel eyes. Even the smells and noises he encountered added to the rich splendor of the Tropics.

Accustomed to the deep grays of Prout's Neck's rough, rocky shores, Homer was completely dazzled by the white sandy shoreline of New Providence. He painted his newly found paradise of sea and sky in rich aquas, vivid emeralds, deep marine blues; adding minute flashes of scarlet and hairline delineations of black.

As a true nature lover he fast acquired an obsession with the palm trees. (They became the featured subject in nearly a dozen paintings.) The long, wind tattered palm fronds were so different from the scrubby huckleberry bushes, junipers, and small leafed trees outside his Maine studio. How his mother would have loved the flora here!

Following footpath trails to the beach or sitting beside coral and shell strewn beaches dotted with ancient cannon, he studied the hemp and Spanish bayonet plants which grew like sentries standing guard. He enjoyed watching the handsome native women walking to market along rough limestone roadways, gracefully balancing heavy baskets of produce atop their heads.

The December sun's warmth was quite surprising and, by then being exceedingly bald, he had to keep his head covered. He felt a new life. As he witnessed new experiences in the Bahamas, and viewed wide, new sea and sky horizons in their shimmering brilliance, he developed a more summary technique for his watercolors. With newfound exuberance, he went about the joy of capturing all that he saw.

As in Maine, Homer's "interest was in human life, out-of-doors under the sky and in sunlight; he liked the simple and unaffected flavors of humanity and showed almost a Dutch genius for the beauty of the ordinary and the everyday."[1] Broadly realistic and boldly free, he transferred onto paper with straightforward accuracy the common daily episodes of Bahamian life.

Homer's initial introduction to Blacks at close range was during his detail duty with the Army of the Potomac during the Civil War. Even then he had been highly intrigued by them, "and he found them of great interest artistically, with their large eyes, high cheekbones, and dusky skin."[2] Their cordiality and quick humor pleased him immensely and he often used them as models and they were most agreeable apparently to pose for his quick pen or brush. Now, in the Bahamas, the local people became his favorite subjects and constant companions. He thoroughly understood them, was interested in them, and sympathetic to their problems. He regarded each individual as important.

Homer also made a number of excursions to the Out Islands and on these trips he painted some of his more monumental watercolors of the Bahamas. Of these, *The Annie Nassau,* now known as *Sponge Fishing, Bahamas,* in which "a white sailing boat filled with Negroes is anchored on the blue bay." The figures, simplified to mere spots of color, are seemingly crowded into the craft. Another boat on the horizon is "almost too detailed and detracts from the effect of the larger one, which it balances quite well . . ."[3]

Glass Windows, Bahamas, (plate 7) is one of the subjects Homer also did while away from New Providence Island. *Native Woman Cooking,* a signed and dated watercolor, bears the notation: "Harbour Island, Bahamas, 1885." Harbor Island is located northeast of Nassau, adjacent to Eleuthera. The tropical watercolor, *Buccaneers,* now in the Cleveland Museum is considered "an historical conception." It demonstrates his advance in his watercolor renderings through "its splendid trees, but he was not at his happiest in romantic imagination, and the painting shows it."[4]

On New Providence and among the Out Islands, Homer witnessed the continual struggle of men against the elements, including the battle for survival between man and shark. This too, became another obsession for him. He studied the shark closely and became completely familiar with the anatomy of this savage of the sea.

While on the boat coming to Nassau in 1884, he'd casually confided to a fellow passenger that he'd like to paint a shark. A few days later a message came from this chance acquaintance that he had a shark for him. Homer, in the meantime, had acquired an oversupply of the creature. Pleasantly enough, according to Beam, he told the messenger, "Tell your friend to go to Hell! I have a dozen sharks!"[5]

Painting thus from Bahamian life and nature, Homer automatically projected much of himself directly into the mind of his potential viewer. He thought not of what he was painting. He merely observed, painted and became transferred psychologically into the minds and eyes of his viewers. His high degree of artistic imagination allowed him to do this, seemingly without effort. "This projection is characteristic of all great art, and of all great artists. The only life their creation possesses is that which is conceived by the union of the minds of the sender and the receiver. The only flame possessed by a genius is that which lights up in the minds of others."[6]

Motivated strongly by his deep understanding and interest in human nature, Homer logically found his best theme in painting the Bahamians occupied with their everyday tasks. This, along with the settings in which he placed his subjects, were undoubtedly considered new and intriguing to the audiences of his day. These scenes, so completely foreign to his critics, were not greeted with immediate acclaim. It was as if they had to grow up to Homer's new technique.

He was apparently prepared for this attitude since Jean Gould stated the following in *Winslow Homer, a Portrait:* "Subconsciously he had suspected that his latest watercolors might not be fully understood; few people could be expected to feel the same excitement that he did over the Nassau pictures. Yet he had hoped that the critics would see in these watercolors a new phase of his work, a firmer technique, a strength that the medium had not attained through anyone else; it still required perfecting, but surely the change must be apparent."[7]

These early pagan watercolors of the West Indies drew less attention than his far paler pictures of Tynemouth. The people who visited the galleries of Gustav Reichard in New York

Market Scene, Nassau (1885), depicts two sailing craft, two rowboats and six natives. They are exchanging a pair of trussed chickens from one boat to another. Privately owned the black and white photo is courtesy of Bradley Olman.

and Doll & Richards in Boston were dismayed by the free and unstudied pictures of bare brown bodies. Sales were few and far between, even at the almost ridiculously low price of $75 per painting.

Sadly, many of the subjects Homer painted on his first visit to the Bahamas have somehow disappeared from public knowledge. Such titles as *Negress with Basket of Fruit, Port of Nassau, Market Boat, Noon, Fox Hill, Banana Tree, Song Birds, Nassau,* and *Near the Queen's Staircase* were mentioned in Downes' biography published in 1911 but they are not museum owned nor are their exact whereabouts known today.

1. E.P. Richardson, *A Short History of Painting in America* (New York: Thomas Y. Crowell Co., 1963), pp. 214-220.

2. Jean Gould, *Winslow Homer A Portrait* (New York: Dodd, Mead & Co., 1962), p. 67.

3. Lois Homer Graham, "An Intimate Glimpse of Winslow Homer's Art," *Vassar Journal of Undergraduate Studies* (New York: May, 1936), p. 8.

4. Philip C. Beam, *Winslow Homer at Prout's Neck* (Boston and Toronto: Little, Brown and Co., 1966), p. 77.

5. Quoted in Beam, p. 169.

6. Quoted from an assessment on creativity by an anonymous advertising executive.

7. Gould, p. 228.

15

An Interesting Cuban Interlude

When Homer learned that Ward Line steamships, which made Nassau one of their ports of call in the West Indies, went next to Guantanamo on Cuba's southeastern coast and then proceeded westward to Santiago, he booked passage on the next vessel.

Shortly after his 49th birthday, the end of February, 1885, he arrived in the ancient Cuban city and a couple of days later wrote to his older brother:

Hotel Lassus
Santiago de Cuba

Dear Charlie

Here I am fixed for a month—(having taken tickets for N.Y. on 8th, leaving 27th of March—this is a red hot place full of soldiers they have just condemned six men to be shot for landing with arms, & from all accounts they deserve it—The first day sketching I was ordered to move on until the crowd dispersed—Now I have a pass from the Mayor "forbidding all agents to interfere with me when following my profession." I expect some fine things—it is certainly the richest field for an artist that I have seen—

Homer had written very casually to Charlie concerning the crowd episode. His attitude of not making a big thing over it was his usual conservative New England outlook. It could have

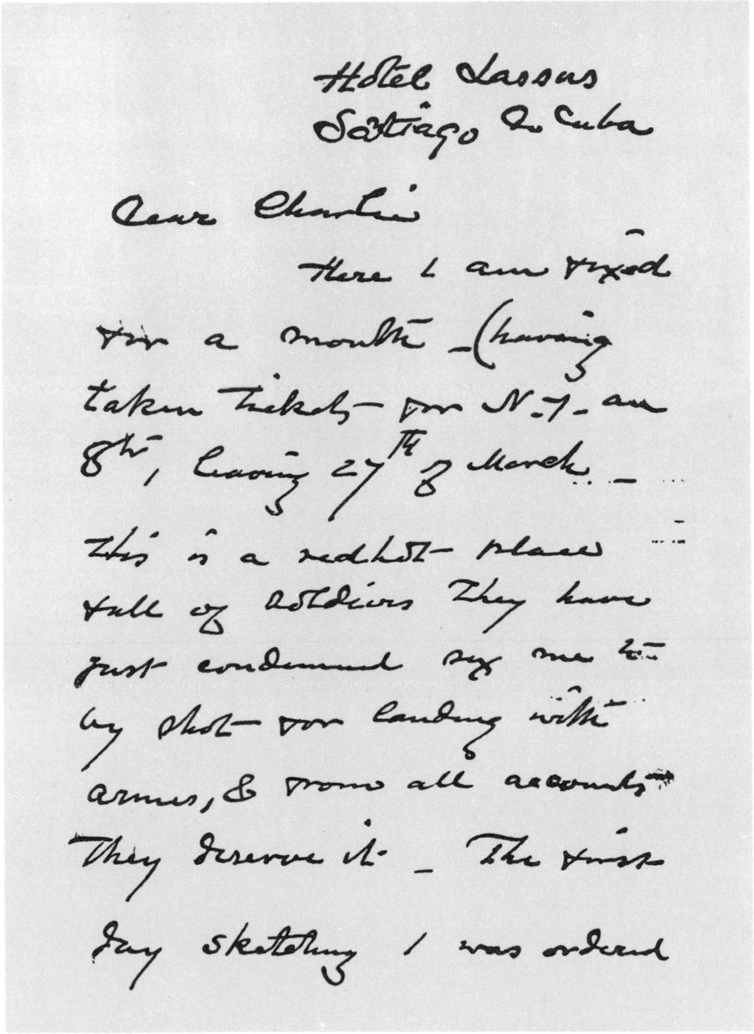
Hotel Lassus
Santiago de Cuba
Dear Charlie
Here I am fixed
for a month — (having
taken tickets for N.Y. on
8th, leaving 27th of March. —
this is a red hot place
full of soldiers they have
just condemned six men to
be shot for landing with
arms, & from all accounts
they deserve it. — The first
day sketching I was ordered

Photo of original letter courtesy of the Homer Collection, Bowdoin College Museum of Art, Brunswick, Maine.

These photos were taken of Señor Luis Manuel de Pando y Sanchez and Señora Cecilia Armand Roch de Pando on their wedding day in 1880. The governor wears the uniform of the provincial commanding general. The photographs were made available through the courtesy of Francisco de Pando, Madrid, Spain, son of Señor and Señora de Pando.

had more serious overtones than he let on. He was a foreigner who spoke no Spanish and when, on the first day, he set up his easel and began sketching at one of the main intersections, a crowd gradually gathered to watch him sketch their familiar narrow avenue.

A Negress, carrying a monkey on her shoulder, was walking on the opposite side of the street. She stopped and stood directly in front of him, so he included both of them in the drawing; other passersby, curious as to the attraction, joined the little knot of people, until it became quite a circle. "The woman with the monkey crossed over and stood directly behind the campstool to get a better look; from its perch the inquisitive creature reached down and plucked at the button on Winslow's cap; he grabbed at the visor just in time to keep the little paws from snatching it off his head. The woman let out a stream of scolding, several others chiming in. A policeman, hearing the hubbub, came to investigate."[1]

Gatherings or meetings were forbidden; crowds were not allowed. In Spanish, the officer brusquely instructed the artist and his audience to move on. There was no alternative for Homer but to pack his gear and go back to his uncomfortable hotel. He walked a number of blocks when he looked up and discovered himself in front of the City Hall. Not one to be easily deterred, he decided to go in, get an interpreter and state his case. The matter was efficiently resolved by the Mayor and Homer could get back to his original sketch.

At the time of Homer's visit, the general and civil governor for the province of Santiago de Cuba was Luis Manuel de Pando y Sanchez who, upon hearing of the street corner incident, sent a message inviting the artist to his palace.

Governor de Pando and his beautiful young wife, Cecilia Armand Roch de Pando, graciously entertained Homer. The Governor brought out his finest rum, distilled from sugar of the Santa Cecilia and Romelia sugar mills, from whose French founders the lovely Señora de Pando was descended.

The de Pandos were honored to have the distinguished American artist as their guest and allowed Homer free reign of their lavish home.

From a balcony, he painted the lacy wrought iron balustrades clearly outlined with the pale Sierra Maestro Mountains in the background. Along the promenade he did a handsome, full-length portrait of the Señora titled *The Governor's Wife* (plate 22), with the traditional black lace mantilla gracefully draped over her head and shoulders. The dark-eyed beauty was only 21-years-old at the time and Homer captured her in a pensive pose

Balcony in Cuba, this watercolor showing heavy tropical clouds hanging over the limestone hills is considered an unfinished study; its chief feature is the intricately wrought fence in the foreground. Photo courtesy of Harry L. Dalton, Charlotte, North Carolina.

standing proudly before a wide horizon of hills and sky. For the first time in a number of years, he utilized again his adeptness for illustrating true feminine beauty. How unlike the stalwart wives of Tynemouth she was; how similar were her dark eyes to those of someone else. Homer had not lost his appreciation for such beauty—merely his own personal desires to have it ever near and to be a possible infringement to his career.

His portfolio was overladen when he boarded his ship to sail home; he was a treasure hunter who had unearthed a rich cache in the West Indies.

1. Jean Gould, *Winslow Homer A Portrait* (New York: Dodd, Mead & Co., 1962), p. 220.

16

Getting Into the Stream of It

On the high seas again, Homer was once more enthralled and exhilarated. As always, wherever he happened to be, subject matter was of utmost importance. With his phenomenal power of concentration, accurate drafting ability, and acute observation, he turned his attention now to the Gulf Stream. He studied its currents and asked questions of those crew members who seemed most aware of its course and educated to its rate of flow.

Homer sensed its tremendous force as he drew practice sketches in his ever present pocket notebook. He envisioned the helplessness of a man alone on a derelict, adrift at the mercy of the Stream's currents with the denizens of the deep surrounding his floundering craft. This vision burned deep. From the sea all around him and its everchanging elements, Homer was uncovering inspiration which would last him throughout his life. He would become the seas' master-painter—or it his master.

The voyage northward lasted 13 days and when his ship docked in New York on Wednesday, April 8th, Homer decided to stay a few days to visit briefly with friends and family before heading on to Prout's Neck. He took care of preliminary arrangements at Reichards for an exhibition later in the fall of the year.

At George W. H. Ritchie's West 14th Street print shop, he stopped in to chat amiably and to obtain a few tips for an etching he'd decided to make of *The Life Line,* from his sea epics series. Ritchie helpfully offered "to put the ground on Homer's plates and send them to him whenever he was ready." Homer's first attempt at etching, *Girl Posing in a Chair,* had been done nine years earlier. This new project would be the first of eight etchings he'd complete from 1885 through 1889. Their titles are *Undertow, Eight Bells, Mending the Tears, Perils of the Sea, A Voice from the Cliffs, Fly Fishing, Saranac Lake,* and *Saved.* Homer had learned about etching from books through careful study of the works of accomplished etchers in Christian Klackner's shop, and from Ritchie. Baker and Cole had also become deeply interested in etching and Gardner ascertained that "it is most probable that Homer was introduced to the art by Cole" in Paris.

The sale of these eight etchings was later arranged for through Ritchie's print shop—but George, "more of an artist and commercial printer than a salesman,"[2] had poor luck in promoting these prints for extensive distribution. Nor was Klackner any more successful in his exclusive franchise efforts of making photogravures of some of these sea pictures and printing of a circular, "Original Etchings by Winslow Homer." Not his favorite medium, etching netted Homer very meager receipts and he eventually gave it up entirely.

While in New York during April 1885, Homer also showed Ritchie some of his Nassau watercolors; one painting in particular of the sponge fishing boats near New Providence Island especially pleased the knowledgeable etcher. (This work was later given to Ritchie by Homer "for helping him on the *'Eight Bells'* etching . . .")[3]

The two craftsmen also discussed which of these tropical works were best suited to line drawing reproduction for one of the leading periodicals.

With his business affairs properly cared for, Homer was able to spend a few days with Charlie and Mattie, telling them of his wonderful West Indies experiences and showing his handsome harvest of art reaped in less than four months.

At last, Homer arrived at Prout's Neck where a high stack of mail awaited his return. His correspondence was never forwarded and might languish six months before he was able to read and answer. He was filled with pride when he again viewed his one-and-a-half story mansard roofed studio house for the first time. This was definitely *home!*

His life that ensuing summer assumed a pattern which it would more or less follow henceforth. He took great pleasure in planting a small flower garden on the house's lee side; later, he trained some of his junipers in Japanese style (a method he'd learned from studying the intriguing dwarfed trees illustrated in Japanese prints and scrolls). To insure his privacy, he constructed a high wooden fence around the garden. In time, he would have an oriental outlook from his studio window.

Painting his house, building a fieldstone wall or supervising the building of a road, busy Homer seemingly became "more occupied with worldly affairs than with his art."[4] His roots were becoming more deeply and solidly embedded in Maine's rocky soil.

The three brothers decided to bolster the Homer real estate holdings at Prout's Neck. They placed their father, Charles Savage Homer, Sr., in charge of the venture thereby reestablishing him as a businessman.

Homer found his best painting time that summer was early morning or late afternoon. In September, after his father had gone to Boston, Homer spent two or more months of concentrated effort completing *The Fog Warning.*[5] The germ for this work had been lodged in his imagination for better than two years. He'd waited painstakingly for months to acquire just the right natural lighting effect.

Homer's initial effort on behalf

of this subject had been sketches of the local handyman, Henry Lee, seated in a dory propped against a sandbank to simulate being pitched by wave action. Henry, dripping wet from a bucket of cold water thrown on him by the artist for authenticity, provided a scowling countenance and his own brand of verbal outburst. From these earliest wash and pencil drawings Homer did a vigorous oil study and titled it *Halibut Fishing.* A photo was made of this which Homer autographed and hung on one wall of his studio. The painting itself was given to a Portland friend. His final large oil, 30″ × 48″, on this theme offers a much greater dramatic impact with the added element of imminent danger of an approaching fogbank. By increasing the apparent distance between the small dory and mother schooner, toward which the fisherman is rowing, more of a test of survival was given. Art critics have long agreed this subject is portrayed with comprehensive realism.

Homer found the hazy light of October just right for another canvas, *The Herring Net,* depicting two fishermen hauling in their catch. A preliminary 1884 drawing titled *A Haul of Herring* had been done with white chalk and charcoal which permitted soft textural effects and broad improvisary quality —ideally suited to the somber mood he wished to convey. (Also closely related to this work and its resulting oil is a watercolor, *Dory, 1884.)*

He also began studies for another canvas, *Lost on the Grand Banks,* but could not proceed due to changing light caused by a heavy fall of snow. The weather grew colder and his provisions at the studio fell low. Mattie and Charlie had invited him to join them in West Townsend for Thanksgiving, but he apparently declined since early in December he wrote his father the following letter from Prout's Neck:

I have just put coal on the fire, which accounts for this smut. I made a mistake in not getting a larger stove. It is very comfortable within 10 feet of it. It heats the room within two feet of the floor, and water freezes anywhere within that space. I wear rubber boots and two pairs of drawers. I know very well what a mistake I am making. I should simply 'irritate my skin & take a cold bath.' But water is scarce. I take a sponge and pick out a certain portion of my body which I do at any time of tide, & always. I break four inches of ice to get at my water. I thank the Lord for this opportunity for reflection, and I am grateful for the advantages I enjoy over Sir John Franklyn.

P.S. Great storm last night. Cold as the d___.[6]

A week or so later he boarded

Sam, his English terrier, with the Libby family for the winter, closed his studio and left for Boston.

Homer did not stay at the Winthrop with his father; he felt it better to room separately at a hotel with less costly accommodations. About a fortnight later he went into New York to visit with Ritchie and to consult with Gustav Reichard concerning the gallery's forthcoming exhibition of his West Indian watercolors. This included fifteen Bahamian and fourteen Cuban subjects, from his previous winter's sojourn to Nassau and Santiago.

The elder Homer complained to his artist son of "beginning to feel rheumatic pain in cold weather" and suggested a trip to Florida. Homer would have perhaps preferred the Bahamas, but "deferred to his father's wishes; he could not afford another trip to Nassau just yet, anyhow."[7]

Their passage from Boston to Key West was probably arranged via the C.H. Mallory & Company Steam Ship Line. This was the first time in many years that his father had taken a sea voyage and Homer found his company and tales of previous experiences enjoyable. While in the Old Island city of Key West, however, he found little time to paint, since—as with some small annoying child—the elderly Homer apparently required continual attention. At 77 years, the father found it delightful to have a doting son all to himself. (One landscape sketch which Homer did paint was titled simply *Key West, Landscape* (plate 27), and upon his return from this trip, he gave it to Charlie and Mattie.)

Another subject, *Under a Palm Tree* (plate 29), mistakenly attributed by Downes to Homer's earlier Nassau watercolor series, was painted at Key West and is so designated, signed and dated January 1886. This thoughtful vertical or upright composition is a watercolor study depicting an attractive mulatto girl clad in a gayly colored dress and wearing a scarf about her head and neck. (In a private collection at present, this work will be in the collection of the National Gallery one day in the future.)

Other watercolors purportedly done on this visit to Key West are *Palms in a Storm; A Norther, Key West; Coconut Palms, Key West; Sand and Sky,* and *Tornado, Key West.* (The last painting mentioned was included in an exhibit held October 16 to November 7, 1915, at the Museum of the Brooklyn Institute. They were loaned with twelve other watercolors by Charlie and Mattie.)

Father and son went next to Tampa, via the Gulf Coast route of the Florida Railway and Navigation Company. They stayed about a month or more at this spot, described by one biographer as "a sleepy inland place near the west coast."[8] Tampa earlier had been an Indian Village and later a Spanish landing site on Tampa Bay. Homer here found ideal fishing in the quiet waters of an inlet. He painted a number of fine watercolors to record his impressions of the wild tropical beauty he found.

Bright in color and warm in feeling with a sense of nature's profusion—the fullness of growth possible in such places, *In a Florida Jungle* (plate 26), could have been painted almost anywhere in the vicinity of Tampa. *At Tampa* (plate 32), with its statue-like renderings of a roseate spoonbill and white egret in a dark, moss hung atmosphere, conveys an almost humid quality. Another subject of this particular trip is *Two Flamingoes, Tampa,* dated 1885.[9]

Live Oaks (plate 31), is initialed W.H. and dated 1886, hence it should be attributed also to Homcr's visit to thc Tampa Bay area where stately, old oak trees, both the live oak and water oak species, grew in abundance. In the foreground center the forms of two hunters and a hound with his tail pointing can be discerned. Long, ghostly streamers of gray moss hang down in the still air.

Spanish Moss; at Tampa (plate 30), depicts the inner bay area with a portion of the peninsula extending into the background above the calm waters upon which sailing craft and a small

Painting the palm trees pulled as taut as bows by the wind, Winslow captured the feeling of a winter storm in the tropics with, *A Norther.* This watercolor is attributed to his visit to Key West in 1886. In a private collection, the black and white photo is courtesy of Hirschl and Adler Galleries, Inc., New York.

steamboat are traveling.

For its 20th annual exhibition in 1887, Charlie loaned *Sketch in Florida* to the American Water Color Society. Downes gave the following description of this 1886 painting: "The crispness of the treatment, the purity and transparency of the color, and the breadth and firmness of the drawing of the palms, palmettos, live oaks hung with Spanish moss, and the other tropical vegetation, were truly characteristic of a master painter."[10]

Thornhill Bar, Florida, another of Homer's works in 1886, is not as popular as some of the artist's other tropicals; it is, however, an excellent example of vegetation of the region. Many portions of the state still offer such richly wild outcroppings.[11]

Indian Hunter Among the Everglades, a vertical watercolor, is also believed to have been painted in 1886. Signed, but not dated, it is similar in technique to *Thornhill Bar, Florida.*

Florida Palms (1886), was lent by Mrs. G.W.H. Ritchie, wife of Homer's late etcher friend, to the Rhode Island School of Design in 1931 for an exhibition.

The summer of 1886, back at Prout's Neck, Homer completed *Undertow* which was later sold to a friend of Charlie's, Edward Adams of New York, for $2,400.[12]

In December, just before the holidays, Homer went to Boston to be with his father. Two months later, the February 1887 issue of *The Century Magazine* went onto the newsstands, carrying nine engravings of his 1884/'85 Nassau watercolor studies. These subjects were created from his initial impressions of Bahamian settings and had been carefully discussed with George Ritchie. Those most suitable to the fine line black and white engraving medium were selected for it. William C. Church's accompanying text on the Bahamas afforded readers an intimate glimpse of the colony as a midwinter resort. Mentioned briefly, is one artist-visitor, "a man of genius and fertile in expedient" who, when it became too cold in Nassau, during the "last winter . . . went to bed in his overcoat. But then he was an artist and the artist's we class with the sensitive plants."[13]

Among the watercolor subjects transferred to engraving was "A Group of Palms" from the original work titled *The Buccaneers* in the collection of Hinman B. Hurlbut, now owned by The Cleveland Museum. Others engraved were "A Hurricane," (Reproduced as *Tornado, Bahamas*—plate 4); *The Conch Divers* (plate 9); *Shark Fishing* (plate 8); and "On Abaco Island," (Retitled *Glass Windows, Bahamas* and reproduced—plate 7).

"A Flower Seller," "A Nassau

Gateway" and "A Peddler" appeared also. The latter sketch later came to bear the title *Negress with Bananas.*

The years 1887 through 1889 found Homer in complete grasp of the subjects presented to him at Prout's Neck. One spring day he wrote to Charlie, "Things are looking very beautiful here today."[14] His letter then went on to banter about his menu and other trivialities, but behind it all was the artist's desire to justify his mode of life which was simple and unassuming in comparison with that of his two brothers.

On summer fishing trips with Charlie in the Adirondacks, Homer found inspiration to carry forward the technique he'd begun in the West Indies. Critics were growing used to his unusual style and had begun to appreciate what he was trying to express. Some even cited his "bold effects of color, extreme breadth and facility of treatment"[15] with unreserved enthusiasm. These watercolors were now considered deserving of the popular attention they received. Homer's art became "admirable examples of painting the essential and omitting the incidental."[16]

1. Albert Ten Eyck Gardner, *Winslow Homer American Artist: His World and His Work* (New York: Clarkson N. Potter, Bramhall House, 1961), p. 180.

2. Jean Gould, *Winslow Homer A Portrait* (New York: Dodd, Mead & Co., 1962), p. 235.

3. M.A.B. *Exhibition Catalog of Water Colors by Winslow Homer,* (Rhode Island School of Design, Providence, Rhode Island, February 6th-March 1st, 1931).

4. Gould, p. 225.

5. Author's Note: *The Fog Warning* was placed on display at Doll Richards' Boston gallery on Friday, January 8, 1886, and was sold almost immediately to Grenville H. Norcross, a local collector and a wealthy cousin of the Homer family. Eight years later Norcross gave the painting to the Museum of Fine Arts, Boston, in the name of the Otis Norcross Fund.

6. Original correspondence in the Homer Collection, Bowdoin College Museum of Art, Brunswick, Maine.

7. & 8. Gould, p. 229.

9. Author's Note: The property of the Museum of the Brooklyn Institute, this watercolor has not been widely exhibited; it was displayed in 1915 with other of the artist's privately owned works.

10. William Howe Downes, *The Life and Works of Winslow Homer* (New York and Boston: Houghton Mifflin Co., 1911), p. 146.

11. Author's Note: This work was acquired by the Museum of Fine Arts, Boston, as a gift from Mrs. Robert B. Osgood.

12. Author's Note: This epic oil painting is now in the Sterling

and Francine Clark Art Institute, Williamstown, Massachusetts.

13. William C. Church, "A Midwinter Resort, with Engravings of Winslow Homer's Water-color Studies in Nassau," *The Century Magazine* (New York: February, 1887), p. 501.

14. Original correspondence in the Homer Collection, Bowdoin College Museum of Art, Brunswick, Maine.

15. Gould, pp. 238-239.

16. Quoted in Gould, p. 239.

17

On the Florida Mainland—Enterprise: 1889-1890.

During the summer and fall of 1889, while deep in the Adirondack woods Homer made plans to go south for the coming winter. Through his brother Charles Homer and his business associates at The Valentine Company, the artist had heard about the small village of Enterprise on the St. John's River in northeastern Florida. This location and its world renowned fishing appealed to him.

There were a number of steamboat lines running from Jacksonville—one of these was the DeBary Merchants Line and the following description of the 200-mile trip is found in an 1885 guidebook:

To ship from Jacksonville (the point of steamboat departure from the upper or lower St. John's River, as you may prefer to term it) is two hundred miles by water, for Sanford, or Enterprise, on that magnificent expansion of the St. John's called Lake Monroe, at the head of large-steamboat navigation. Steamers leave their wharf at half past one P.M., reaching Palatka early the same evening, and Enterprise and Sanford about noon next day, where excellent hotels are found.[1]

The round trip fare was ten dollars. It was also stated that a large hotel "to eclipse anything of the kind in the South as a winter resort, is being constructed there, for which the locality is peculiarly adapted."[2]

To the artist, the St. John's River, actually a tidal estuary of the Atlantic Ocean, offered a picturesque series of lagoons and lakes, unbelievably abundant with fish.

Along the shores of the river stately homes had been built. One of these mansions was that of Frederick DeBary who first came to America in 1840 as an agent for Mumm's Champagne. His handsome white columned house was built of sturdy cypress and served DeBary and his family as a hunting and fishing preserve. Here, he lavishly entertained many notables, including Grover Cleveland, U.S. Grant, the Prince of Wales—who later reigned as Edward VII, Harriet Beecher Stowe and DeBary's own New York neighbors, the Astors and the Vanderbilts.

The overall area of Lake Monroe had recently become popular to both wealthy Americans and the aristocratic tourists from Europe. Its recreational facilities were outstanding. The twice-weekly arrival of the Brock Line steamboats at the pier of the area's largest hotel, the Brock House, was a gala occasion. The steamboat, *Florence,* ran daily (except Sunday) from Jacksonville to Palatka, and the *Darlington* and *Hattie* left Jacksonville for Enterprise on Wednesdays and Sundays. These Brock

A photograph of the artist taken with his own camera during his trip to Enterprise 1889-1890. Homer has turned slightly in the skiff's stern while his two black guides row across the calm waters of Lake Monroe on the St. John's River. The artist's meticulous attire of long-sleeved white shirt and vest with straw hat is typical. Photo courtesy of the Homer Collection, Bowdoin College Museum of Art, Brunswick, Maine.

Line ships were quite splendid; "built for passenger trade, with luxurious staterooms and fine appointed salons." The cuisine served was excellent and famous with "venison, wild turkey, other game, fish, and fruits, all native to the country and in seemingly endless abundance." The line's advertising explicitly stated: "The Old Reputation of this Popular Line will be fully sustained, and every Comfort Guaranteed to its Patrons."[3]

Jacob Brock's 100-room luxury accommodation, erected in the palmetto jungles, lured many wealthy personalities, but Homer's aversion to snobbery more than likely led him to choose instead to stay at the Lemon House or Live Oak House—which were also $2 per day cheaper.

There were approximately 1,000 persons living in Enterprise in the late 1880's and the community boasted having a recently erected house of God, All Saints Episcopal Church.

The St. John's River was home for gigantic bigmouthed bass and, as an avid angler, Homer found this the greatest sport of all. The artist dearly loved to fish: he liked their colors, the shining silvers and reds, and loved the action they provided, their courage and fighting strength.

The familiar watercolor, *Bass Fishing, Florida,* was completed in the winter of 1889/'90 along with other airy, luminous, light-suffused St. John's River landscapes.

One watercolor of this season's output was *Blue Spring, Fla.* and it presents a curious bit of discrepancy since there was a favorite swimming hole called Green Springs, located a few yards from a mysterious sinkhole on the property of Dr. James Stark, one of the area's most prominent citizens.

Downes in his biography of 1911 seemed to feel that Charlie accompanied his brother on this trip to Enterprise and continued his remarks by writing:

> *I have seen a series of photographs taken by Winslow Homer on this occasion, and in one of the snap-shots there is a good sized wooden box in the foreground, which was used to carry luncheon in. It chanced to be an old whiskey box, and the name of the distillery was stenciled on the side of the box. When Winslow Homer came to develop the print, and to give his brother a copy of it, he pointed out the lettering on the box, and said, with a smile, 'This sort of thing is calculated to give people a wrong impression.'*[4]

Downes also incorrectly stated, "Enterprise is a little-known winter resort;"[5] but then, perhaps by his time such places as Palm Beach and St. Petersburg were better qualified to

Identified as a Florida setting, it can be surmised that this photo taken by Homer shows the lodging in which he stayed while at Enterprise in 1889-1890. It is courtesy of the Homer Collection, Bowdoin College Museum of Art, Brunswick, Maine.

Blue Spring, Florida is one of the watercolors attributed to Homer's trip to Enterprise, Florida in 1889-1890. Privately owned the photo is courtesy of Bradley Olman.

lure tourists and the little community in Volusia County sank into oblivion, except for the wonderful renderings Homer left.

Further comments concerning photography are afforded by Philip C. Beam, who apprizes that "even a cursory comparison of any of his paintings with the similar photograph is an illuminating comment of the extent of his imagination and departure from actuality in creating these works that we commonly consider the ultimate in realism."[6]

One such comparison is available in the photo "Palm Trees, Florida" and the watercolor, *Palm Trees, St. John's River.* (plate 38)

It should further be recalled that during the Civil War, while he was depicting scenes of battle for *Harper's Weekly,* Homer probably met the well-known photographer, Mathew Brady, and his assistants, Alexander Gardner and Timothy O'Sullivan, and, in their works, saw "the striking emphasis on silhouetted forms,"[7] a technique used most effectively in the above mentioned watercolor.

This camera, which Homer used to produce photographs that significantly paralleled the subject matter of his tropical watercolors, as well as those works of the Adirondacks, was an Eastman Kodak No. 1. "An advanced model for its time, it took one hundred pictures two and a quarter inches in diameter on roll film and was an excellent instrument for a traveling artist." This equipment was introduced in Minneapolis, Minn., at an 1888 photographer's convention. Homer's "extensive use of a camera shortly thereafter and throughout the '90's on his own trips shows that he was not only interested in photography, but alert to new developments."[8]

He proved, however, that "he knew the difference between a photographic record and an interpretive work of art, and he never confused the capacity of the camera to record details with extraordinary clarity (which misled many weaker painters) with the artist's need to view selectivity."[9]

Fishing and painting, Homer produced at Enterprise a series of studies of game fish which, because of his "intimate acquaintance with his subject,"[10] are considered unique in American art. One of these works, *Trout Breaking* (1889) in the collection of the Museum of Fine Arts, Boston, accurately depicts the life of these game fish in the St. John's River. "Strikingly similar and brilliant" is *Bass Fishing, Florida,* in which Homer "intensified the luminosity of the reflections from the water in a way that the camera (at least of that day) could not match."[11]

Two Trout (1889), in the collec-

"Palm Trees, Florida" photographed by Homer with his circular Eastman Kodak camera during his fishing trip to the St. John's River in 1889-1890. It is very like his watercolor, *Palm Trees, St. John's River, Florida* (plate 38). Photo courtesy of the Homer Collection, Bowdoin College Museum of Art, Brunswick, Maine.

Three Men in a Boat, a companion subject to *Rowing Homeward* (plate 35), points up Homer's interest in reflections. According to one cataloguer, "the whole weight of the work rests on a wash technique . . . the men and their rowboat are almost as fluid as the reflections . . ." The photo is courtesy of Colby College Art Museum, Waterville, Maine.

tion of IBM Corporation, is a watercolor whose location has been questioned. Some feel it was done in the Adirondacks, others prefer to attribute it to the Enterprise studies; a still life, oriental in concept, this subject was among the works shown at the Century Loan Exhibit at Prout's Neck in 1936.

We should consider carefully in viewing these subjects that in a sense Homer thought not of what he was painting, but of the impression his finished work would leave on the minds of persons receiving his message. Psychologically, he transferred himself wholly into the minds and eyes of his viewers. To do this he had to possess a high degree of imagination. How many viewers have previously given thought to Homer's especially high interest in the mental reaction of others? Was he perhaps in his own way by depicting the wilderness areas of America's tropical Florida as accurately as possible an early conservationist? Various writings have stated that though he traveled extensively with hunters, he never shot and killed an animal. By reproducing the virgin beauty of these places was he attempting to forewarn that changes were to come? Luckily, Enterprise is one quiet little spot which in many ways still offers the serenity and idyllic scenery Homer depicted.

1. G.M. Barbour, *Florida for Tourists, Invalids & Settlers* (Gainsville: University of Florida Press, facsimile reproduction 1964), p. 192.

2. Ibid., p. 121.

3. Edith G. Brooks, *Saga of Baron Frederick de Bary & de Bary Hall, Florida* (Florida: Convention Press, Inc., 1966), p. 10.

4. & 5. William Howe Downes, *The Life and Works of Winslow Homer* (New York and Boston: Houghton Mifflin Co., 1911), p. 136.

6. Philip C. Beam, *Winslow Homer at Prout's Neck* (Boston and Toronto: Little, Brown and Co., 1966), p. 75.

7. John Wilmerding, *Winslow Homer* (New York, Washington, and London: Praeger Publishers, 1972), p. 43.

8. & 9. Beam, p. 75.

10. Philip C. Beam, "Winslow Homer at Prout's Neck" *Catalog for Bowdoin College Museum of Art* (Massachusetts: 1966 Exhibition), p. 12.

11. Beam, p. 87. (Author's Note: Beam terms Homer the most ardent of fishermen with a passion for the sport, "his keen study of the habits of various species, and his admiration for their fighting qualities made him an authority on the subject . . .").

18

Presenting Nature's Overall Tapestry: 1890-1897

Winslow Homer was a perennial student of nature and his major works basically deal with the presentation of the flora, fauna, sea, sky and terra firma. As he matured in his craft, Homer was not one to theorize on art. Rather, he took up his palette, set up his easel and painted exactly what Mother Nature presented to him. If it happened to be calm seas, or flora and fauna, or the verdant jungle growth of the tropics he accepted it.

Just as his semi-contemporary, George Inness, was "addicted to meadows and worried by elaboration,"[1] so Homer was drugged by the sea and ultimately depicted it with severe simplicity. He thoroughly "enjoyed the roar and riot of wild waters . . . the more the surf crashed, the more he reveled in it." Areas, like Tynemouth and Prout's Neck, frequently "haunted by cataclysmic storms,"[2] were Homer's favorite places. When nature raged, his inner spirit surged, spurring him into the creation of soul inspired works that meant much in increasing his stature as one of America's greatest artists.

Notable examples of his rendering the sea's dramatic moments are found in these oil paintings of the winter of 1890-1891; *Sunlight on the Coast* received highly complimentary press notices after its showing in 1891 in New York. (It is now in the collection of the Toledo Museum of Art.) *Winter Coast,* depicts the great desolation and wintry chill of the Maine coast when the bone penetrating cold pervades. (It is now in the Philadelphia Museum of Art, John G. Johnson collection.) *West Wind* represents the epic grandeur of a shrieking autumn gale with 60-foot waves and the winds "whipping flying mares' tails from the crests of the towering breakers, and at precisely the most effective spot is the figure of a woman struggling along the marginal way that runs by the foot of High Cliff." Homer, satisfied with this brilliantly created study in browns, termed the work, "Damned good!"[3] He also won a $100 wager with John LaFarge who had criticized Homer for his dull-toned paintings and frequent use of brown. Reichard displayed the work and the public and critics alike accepted and admired it.[4]

During the summer and fall months of 1891, Homer was in northern New York, staying at the clubhouse of the Adirondacks Preserve Association. One of the paintings he completed on this sojourn, *Guide Carrying a Deer,* was given to Charlie for Christmas, 1891. He and Mattie had been to Europe during that summer and Homer was most anxious that Charlie's eye, "fresh from European pictures,"[5] would be proud of his brother. This watercolor was a variation of an oil subject, *Huntsman and Dogs,* for which Homer used 14-year-old Wiley Gatchell as the model. The hounds in this picture had been specially purchased by the artist and when he was through with them, he gave them to a neighbor–there would be no rival for Sam, his English terrier.

Other of his Adirondack subjects are *The Mink Pond* (1891), which beautifully offers a bit of Homer's touch for humor with a frog and water lily squared off "in an amusing juxtaposition"[6]–balanced on either side of the paper. Similarly, *Leaping Trout* (1891) treats us to trout cavorting in midair above the same pond with its lily pad leaves. Here one realizes the cool, sure "organizing power of a distinguished and unusual mentality." Gardner sees Homer as being nature's own artist, always an "accurate reporter, the scientific observer."[7]

In almost a photographic sense, *Trout Breaking* (1889) has playful contrast with the jumping trout adding a pair of butterflies hovering nearby. The artist's sharp sense of humor–which often extended to practical jokes on the family on some occasions–is most evident in his flora and fauna subjects. One pertinent example of this appears later in his study, *Bermuda Settlers* (plate 60). Homer laughed frequently at his pictures, his family, and himself. He relished a good gag and quite frequently pulled them on his father to keep the old man's spirits up in his later years. Homer's letters were continually spiced up with humorous sketches of his life at Prout's Neck. And, ordinarily very sensitive about his baldness, he even laughed and joked about it among close friends.

The mid to late 1890's found Homer producing such great paintings as *Maine Coast* (1895) and *Northeaster* (1895) for which he was awarded the Temple Gold Medal in 1902 by the Pennsylvania Academy of Fine Arts. Later came *Watching the Breakers: A High Sea* (1896) and *Weather Beaten* (1897), which also bore the title *Storm Beaten* for some time. *Cannon Rock* (1895) won for Homer another gold medal, this one from an exhibition in Charleston, South Carolina.[8]

Beam assesses that at Prout's Neck Homer "gave American seascape painting a wholly new dimension by ignoring the mild-mannered panoramas of his predecessors and carrying us with him to the ocean's edge . . ."[9] Interestingly too, W.A. Rogers in his 1922 volume, *A World Worth While,* stated that before the 1870's "there was little real effort on the part of American illustrators to interpret nature" and went on to write: "I verily believe, however, that the first great impulse in the new American art of illustration came from Winslow Homer." It was Homer's works which Rogers felt "made a deep and lasting impression"[10] on other artists of his period.

Wild Geese painted at Prout's Neck in 1897 was first exhibited at the Carnegie Institute, Pittsburgh, with the title *Flight of Wild Geese*. Downes assessed this canvas as exemplifying "the originality of the artist's observation and his extraordinary instinct for a fine composition. The file of startled wild geese flying above the sand dunes, where a pair of their unfortunate fellow-fowls have just been brought to earth by a shot, is remarkable in its swift movement and the pattern of the picture is extremely interesting."[11] The first owner of this subject, the late Mrs. Roland C. Lincoln, Boston, claimed its initial title was *At the Foot of the Lighthouse* and that the geese had not been shot, but were killed accidentally, "when they flew into the structure, a not uncommon occurrence."[12]

The Fox Hunt (1893) Beam feels demonstrates how well Homer "assimilated the essential principles of oriental design—precision, lucidity and orderliness combined with narrative comment—without sacrificing his personal manner of expression."[13] This fauna subject was painted by Homer from a dead fox and dead crows which he had a neighbor at Prout's Neck secure for him; arranging the birds "as if in flight on a snowdrift outside the studio window," he let them freeze stiff. The fox was posed in a "running attitude"[14] by means of sticks and string. Unfortunately, the painting was begun rather late in the winter season and when the weather grew warm the crows thawed. He completed it by luring live crows with corn.

Homer went to great pains for authenticity in many of his paintings of terra firma also. His rock formations recaptured in oil and watercolor depict practically every well-known grouping to be found along Maine's coast in the vicinity of Prout's Neck. He would wait hours for just the right light to appear and paint for mere minutes trying to recreate an effect. He was fascinated by the strata to be found in the Bahamas and Bermuda and records his impressions in *Natural Bridge, Coral Formation* (plate 59), and *Rocky Shore, Bermuda*.

It is as if he considered it his role as an artist to constantly seek out instances where his talents were most effective in presenting all of Mother Nature's overall tapestry for the world to enjoy. Sadly, most of these works were completed after his mother had passed away; she never knew how strongly her interest in natural beauty had influenced her son.

1. James Thomas Flexner, *The World of Winslow Homer 1836-1910* (New York: Time Incorporated, 1966), p. 97.

2. William Howe Downes, *The Life and Works of Winslow Homer* (New York and Boston: Houghton Mifflin Co., 1911), pp. 9-10.

3. Philip C. Beam, *Winslow Homer at Prout's Neck* (Boston and Toronto: Little, Brown and Co., 1966), pp. 96-98.

4. Author's Note: This handsome oil is in the collection of the Addison Gallery of American Art, Andover, Massachusetts.

5. Jean Gould, *Winslow Homer A Portrait* (New York: Dodd, Mead & Co., 1962), p. 244. (Author's Note: This is quoted from a letter written to his brother on October 15, 1891.) (See also—Beam, p. 99).

6. John Wilmerding, *Winslow Homer* (New York, Washington, and London: Praeger Publishers, 1972), pp. 169 and 179. (Author's Note: This painting is in the Fogg Art Museum collection.).

7. Albert Ten Eyck Gardner, *Winslow Homer American Artist: His World and His Work* (New York: Clarkson N. Potter, Bramhall House, 1961), p. 226.

8. Author's Note: This famous painting depicts a well-known rock formation just a few hundred yards from Homer's Prout's Neck studio; the artist, however, altered the placement slightly for aesthetic balance.

9. Beam, p. 139.

10. Quoted in Gardner, p. 157.

11. Downes, p. 201.

12. & 13. Beam, p. 248.

14. Ibid., p. 109.

19

Returning to the Bahamas: 1898-1899

In the later 1890's there were many influences to account for Homer's solitary life pattern and periods of depression. Though they had not kept close contact with each other in recent times, Homer must have felt some sadness upon hearing of the passing of his old friend, Joseph Foxcroft Cole, in 1892.

Gone in fact, were many of the older male residents who had spent their entire lifetimes at Prout's Neck—men with whom Homer was on good terms. This void also created a lack of suitable models of the particular type character he enjoyed painting—the rugged, bearded and weathered fishermen of Maine, who often resembled the guides he'd used as models over the years in the Adirondacks.

The artist had long understood his own need for a retreat, his deep-seated requirement for a private introspective world. He had become a master craftsman with his solitude being of infinite importance. Now, suddenly it seemed to him this very solitude had become a vicious circle, engulfing him and making him restless again. He had found "a kind of wildness" at Prout's Neck "which appealed to a deep romantic streak" in his nature, "making him feel at home with lonely scenes which other men avoided."[1]

Then, during the winter of 1897, Homer's beloved English terrier, Sam, died. Dog and master for 16 years had been familiar figures to the Neck's residents as the pair scrambled along the rocky, wave splashed shoreline. In later years, the dog had grown very fat and the long daily walk to the post office with his master had become an arduous journey for him. This meant stopping frequently, coming and going, and to let the children say hello to Sam.

Homer truly mourned for his dear old pet, but he wanted no other. Charlie, upon receiving word of the dog's passing, wrote a letter suggesting that he could send his brother an English bull pup—not only for the companionship, but also as a watchdog since Homer's studio had been broken into a couple of times and a number of his gold medals and other mementos had been stolen. The artist's letter of reply to this offer stopped this action:

January 14, 1897.

On no account send me a dog—the only companion I want is a Bobolink (bird) & the next time I go to Boston I shall get one . . . As for robbers, I have no fear of them, sleeping or walking. I am a dead shot & should shoot, without asking any questions—[2]

(Later while visiting his father, in Boston he went to Ludlams'—importers and dealers in fine singing canaries and talking parrots, and purchased a pair of birds for himself.)[3]

Nor was Homer always sure whether he really cared to continue painting. Seemingly, he "did not care whether he painted or not" according to his old Bufford's associate, Joseph Baker. Homer wrote, at present, and for sometime past, I see no reason why I should paint any pictures.[4]

Part of his baffling remarks and reticent attitude about painting may be attributed to his father's failing health and growing cantankerousness. The elder Homer died August 22, 1898, at the age of eighty-nine.

With the old man gone, the *Ark* seemed more lonely than ever; for fourteen years it had been without its mistress, Henrietta Homer, and now with the patriarch gone, the huge frame house represented past pleasures which could not be retrieved.

The family decided, in complete agreement, that Charlie and Mattie should take over the *Ark*. This meant a thorough redecorating job and the addition of an outside chimney and a Richardson Romanesque tower, providing a billiards room for Charlie and a tearoom for Mattie. Arthur and Winslow, since the beginning of developing Prout's Neck as a summer resort, had argued about property lines of lots they owned adjacently. As the value of this land increased, their arguments grew more frequent. Finally, in September 1898, they tossed a coin to decide which of them would acquire "the front lots and which the back on Eastern Point." Losing the toss, Homer got the back lots along with a front "half of a triangular piece of land outside the marginal way."[5] From this vantage point, one of his most famous Maine works, *Eastern Point, Prout's Neck,* was painted.

In many respects, this was a winter of re-discovery for the artist; at this time he left for the Bahamas and there found a renewed enthusiasm which enabled him to paint more surely and strongly than ever before.

Lloyd Goodrich claimed that many self made artists are slow in reaching maturity. But it should also be pointed out that Homer's intensity and concentration with the creation of paintings was unlike some of his contemporaries. LaFarge, Whistler and other artists of his period were men of various careers—authors, teachers, decorators and designers of stained glass—dividing and diffusing their talents. Homer stuck diligently to his field—diversifying from medium to medium, where feasible, as training, talent and technique advanced. He was now approaching his 62nd birthday and about to present the world a new scope of subjects in "sunlit waters," fully imparting "his enthusiasm for nature" so that we "seem to be with him on the spot."[6]

For several winters he had wanted to return to the West Indies but had "changed the plans to suit his father's preference." Nothing stopped him this time and he was truly delighted to find that everything seemed quite the same to him as it had in 1884. He found it all wonderful. He was working with the same "zest and raciness of the carefree black boys who were again his favorite subjects—he gloried in painting these pictures of gladsome youth, unfettered by the bonds of civilized society."[7]

Each morning Homer was up and out early to rapidly paint the brilliant pageantry. He found a new daring in his coloration with bold sweeps of emerald above indigo. There was a feeling of ecstacy and it was puzzling—yet he also felt a rush of happiness and realized that he was finally attaining "his true contribution to the world of art." Inwardly, Homer was realizing a "deep satisfaction of discovery all he had come back to find."[8]

Making the utmost of every moment, he painted with restored vigor. Gone were the doubts of his own ability to convey the messages he witnessed, every brush stroke was seemingly inspired.

Ship Letter
We expect to arrive tomorrow morning & meet the Sister ship from ~~S~~ Santiago on the way to New York — I have had a fine voyage

Ship letter by Winslow Homer. The date on the following page in his diary-notebook is December 10, 1898. It was probably written while on board his ship enroute to Nassau. Photo of original letter courtesy of Homer Collection, Bowdoin College Museum of Art, Brunswick, Maine.

The results of this sojourn were immediately evident, not only to him but to the staff at Knoedler's, the gallery he had engaged following the retirement of Gustav Reichard in 1897. It was to Charles R. Henschel, "who was inclined to favor his oil paintings," that Homer made the brash prediction when he was at M. Knoedler & Co. arranging for an exhibition: "You will see, in the future I will live by my watercolors."[9] These latter Bahamian studies were, indeed, some of the artist's strongest watercolors.

Leaving the West Indies around the middle of February, he wrote to Thomas B. Clarke, the artist's Boston patron who eventually owned more of his works than any other collector:

I have had a most successful winter in Nassau. Found what I wanted, and have many things to work up into two paintings that I have in mind. I shall not go north until it is warmer, but am through working for the winter, and desire to report myself very well.

The letter was dated February 25th, the day after his birthday, and was probably written from Florida.[10]

The many things Homer wrote of included: *Under The Coco Palm,* dated 1898 (plate 41), which with its "twin shapes of the coconut and the boy's head"—centered in "an effusion of colorful flowers and leaves," characteristically conveys "felt warmth and implied well-being."[11]

Sloop, Nassau, dated 1899 (plate 50), another of that winter's creative output is described as a "vivid concentration of form and coloring, strong simplicity of design, and utter control of medium . . . the furled sails and their bright reflections in the water create the central focus of his design. The curving furls possess an abstract rhythm that is fixed against the dark blue sky behind them. By working from the bare white paper in the center to the dark washes at the side, Homer establishes a vivid range of tonal contrasts. His brushwork is free and sure, the strokes of wash (are) broad, swift, and expressive. It is remarkable that he should so readily achieve an effect at once of ease and drama."[12]

Donelson Hoopes considers *Hurricane, Bahamas,* dated 1898 (plate 44), "a curious picture, at once strangely awkward and yet convincing. One has the impression that it was painted directly from nature, perhaps very quickly during the interval between the gathering of the storm's first clouds and the onset of the torrential winds and rain. There are discrepancies in the drawing that seem purposeful: the palms are indicated with ease and infinite sensitivity, yet the houses seem flat and crude in execution. The flattened perspective of the buildings again offers a reminder of oriental approaches to drawing."[13]

Hoopes further goes on to mention Beam's comparison in the treatment of the palms to the manner in which the 14th century artist, Wu Chen, depicted bamboo during the Yuan Dynasty. "Whether or not such comparisons actually reflect Homer's inspiration for this watercolor, the work is, undeniably, one of his most powerful statements. The grouping of low triangles representing the gable ends and roofs of buildings suggest a certain tenacity, as if they were living things huddled and clinging to the earth lest they be swept away in the fury of the storm; while the palms, in spite of their apparent fragility, stand erect, defving the power of the hurricane."[14]

Homer, in treating his watercolor subjects, whether they were of the tropics or elsewhere, had long ago learned not to be too glib or facile in his handling; his "touch is the natural and consequent outcome of the theme. It is largely the design which gives them their potency in memory—they have their stamp, and this is indeed their cachet . . ."[15]

Continuing Downes' summation: "Nature, broad, spacious, elemental, seems to have sunk into his mind, fixed there in some shape or pattern, strong spots of sky or water, almost savage at times in their coloring. His vision is as clear as a

Sailboat at Nassau, a signed and dated watercolor, was completed by Homer in February of 1899 during his second visit to the Bahamas. One interesting and unusual technique the artist used was to scrape the hull and sails down to the bare paper through the wet surface. In a private collection, the black and white photo is courtesy of Wildenstein and Company, New York.

window pane through which one might look out upon the scene; but the selection is that of an artist who seizes the most salient and typical point of view. There is no softening of effect nor prettifying of facts—great nature suffices, and his works possess the true beauty of essential fidelity. Design is always there, for it is the mysterious power of unerring choice. This it is which places Homer above the plane of a competent painter and proclaims him an artist of the first rank."[16]

While Homer was in the Bahamas for the winter season of 1898-1899, Thomas B. Clarke put his extensive art collection up for sale; included were thirty-one of Homer's paintings—sixteen oils and fifteen watercolors. Writing to his good friend and patron, Homer said:

I have only just received news of the great success of your sale. I owe it to you to express to you my sincere thanks for the great benefit that I have received from your encouragement of my work, and to congratulate you. . . . Only think of my being alive with a reputation (that you have made for me).[17]

Although Homer did not directly benefit from Clarke's sale, the prestige of such a sale was enormous. His paintings sold for the sixth, seventh, and eighth highest prices, which was an indication of the value Clarke—and the public—set on them.

Homer's oils brought $30,330 and the watercolors, $2,965 for a grand total of $33,295. The sale definitely raised the value of other of the artist's works and shortly afterward twenty-seven Canadian watercolors were exhibited by Knoedler's and fourteen of these sold immediately at prices of $200 to $250 each.[18]

According to Beam, the Clarke auction enhanced the artist's reputation "among the general public, which, then as now, often measures artistic worth by monetary value."[19] Homer found this also another stimulant to continue his painting. Truly, his second visit to the Bahamas had been a most victorious happening.

1. Philip C. Beam, "Winslow Homer at Prout's Neck" *Catalog for Bowdoin College Museum of Art* (Massachusetts: 1966 Exhibition), p. 7.

2. Original correspondence in the Homer Collection, Bowdoin College Museum of Art, Brunswick, Maine.

3. Jean Gould, *Winslow Homer A Portrait* (New York: Dodd, Mead & Co., 1962), p. 50.

4. Quoted in William Howe Downes, *The Life and Works of Winslow Homer* (New York and Boston: Houghton Mifflin Co., 1911), p. 167.

5. Philip C. Beam, *Winslow Homer at Prout's Neck* (Boston and Toronto: Little, Brown and Co., 1966), p. 158.

6. Donelson F. Hoopes, *Winslow Homer Watercolors* (New York: Watson-Guptill Publications, 1969; third printing 1971), p. 42.

7. & 8. Gould, pp. 264-267.

9. Quoted in numerous works, [See–James Thomas Flexner, *The World of Winslow Homer 1836-1910* (New York: Time Incorporated, 1966), p. 120].

10. Quoted in Gould, p. 268.

11. & 12. John Wilmerding, *Winslow Homer* (New York, Washington, and London: Praeger Publishers, 1972), p. 169.

13. & 14. Hoopes, p. 44.

15. & 16. Downes, p. 97.

17. Quoted in Gould, p. 268.

18. Downes, pp. 204-205.

19. Beam, p. 159.

20

A New Place of Discovery: Bermuda 1899-1901

Winslow Homer went to Bermuda for the first time in December, 1899. Because of the island's proximity to the Gulf Stream, average daily temperatures during the winter season ranged from 65 to 70 degrees; a climatic factor which the artist thoroughly favored.

The artist found no trouble in locating a small, comfortable bungalow for himself, at a reasonable price for the season. Roaming throughout the northeasterly section of the island he located many ideal settings for his watercolors.

In evaluating his Bermuda subjects, one modern day exhibition cataloguer feels they may be considered to have two pictorial emphases: the first being "hillside houses with profusely flowering gardens overlooking a calm bay" which in a sense contains an impressionistic quality—the foliage being defined by color rather than by line, as evidenced in *Bermuda,* dated 1900 (plate 57).

This group, which includes *Flower Garden and Bungalow, Bermuda,* dated 1899 (plate 55); *Shore at Bermuda* (plate 54); *Salt Kettle, Bermuda;* and *Inland Water, Bermuda* is also considered related to such watercolors as *A Wall, Nassau,* dated Dec. 31, 1898 (plate 47), Homer's Bahamian subject of the previous year.

The second group has themes of limestone cliffs facing open water and includes a kinship with an earlier Bahamian study, *Glass Windows, Bahamas,* dated 1885 (plate 7). The most important painting in this category is *Rocky Shore, Bermuda,* dated 1900 which is characterized by the "subtleties of line." And, in this particular work, "the eye is thrust out in perspective to the horizon and brought gently back by an increasing wave motion."[1]

Also fitting into this classification are *Natural Bridge* and *Coral Formations* dated 1901 (plate 59). Both of these works have previously been classified as belonging to Homer's Bahamian watercolor series, but careful study has inclined this writer to think differently. Along the southern coast of St. George's province, below Tucker's Town, is a group of naturally sculpted towering rocks designated on current maps as Natural Arches.[2] It is possible that these two watercolors and *Rocky Shore, Bermuda* were created in this vicinity.

Donelson Hoopes offered this data on *Natural Bridge:* "As much as Homer's studies of Bahamian walls would seem to indicate a playing with notions of occult balance, so this aerial view of the great natural bridge signals a release from the expected composition of a landscape view. A large portion of the paper is given only the slightest articulation, and Homer seems to enjoy letting stand that long, unbroken line of the horizon—

The Coming Storm, painted in 1901, is a brilliant watercolor. It is termed by Beam to be, "almost the peak of his work in that medium." He describes it as being, "highly simplified, . . . every device of technique, tonal organization, and design strike the eye with maximum visual impact." Privately owned the black and white photo is courtesy of Bradley Olman.

against all conventional rules of good compositional form. As the eye draws near the bridge, the tempo quickens until, at last, all the accents of color erupt in the lower right corner in a nearly Fauve brilliance, dominated by the soldier's scarlet tunic. Such pictures tell much about the artist's capacity for creative invention. For Homer, it suggests that had he not been so convinced of the virtue of painting as closely as possible to nature, he might have extended his enormous capacity for invention beyond visual reality."[3]

Rocky Shore, Bermuda depicts a bright blue sea paling to blue green between gray cliffs; two redcoated soldiers on a white beach are at the right. The work is signed and dated lower left "Homer 1900." The subject has also been called "Scene in Nassau," but is inscribed on the back in the artist's hand "Homer—Landing Bermuda—1900."

The artist evidently found life very pleasant in the Bermudas for only on one occasion is a more somber note detected. The subject is *The Coming Storm,* dated 1901, which was acquired by Homer's dealer, Charles R. Henschel of M. Knoedler and Company for his own personal collection. It was one of Henschel's favorites among his extensive array of Homer's works.[4]

A more specific location is represented in *Salt Kettle, Bermuda* which Homer created along Salt Kettle Road looking nearly due north across Hamilton Harbour toward the province of Pembroke. Today there are a number of small guest houses at this location, providing meals and accommodations for vacationers.

Bermuda (plate 57), depicts a northern variety of the sea grape plant which does not have quite as large leaves as those species growing wild in the Bahamas and other southerly climates.

Another watercolor, *View from Prospect Hill, Bermuda,* was painted in a location northeasterly of Hamilton in the vicinity of a military housing area where British troops were billeted. The picturesque island of Bermuda has many parishes offering roads with uniquely descriptive names; i.e., Happy Valley Road, Marsh Folly Road—overlooking Pembroke Marsh, and Dock Hill Road—leading down to Devonshire Dock.

The artist seems to have made an error in titling *Gallow's Island, Bermuda,* a fine watercolor which he painted along North Shore Road in Hamilton Province, looking westerly toward the open sea beyond Shelly Bay Shoal. The correct name of this small island is Gibbet Island, which according to Webster means a gallows or a structure like a gallows from which bodies of criminals already executed were hung and exposed to public scorn. The island was very possibly Ber-

Salt Kettle, Bermuda, depicts reflections in a placid cove of inland water. Now in a private collection, the subject was in the possession of the artist's brother when a Memorial Exhibit was held at the Metropolitan Museum of Art in 1911. Privately owned the black and white photo is courtesy of Bradley Olman.

muda's "hanging place" in the early days, before British law and order prevailed.[5]

There were other Bermuda creations and among these are *In a Paget Garden, Distant View of Great Sound,* which Homer could very well have painted in the province of Paget, looking westerly. (This is in the vicinity of Salt Kettle peninsula.) Another title is *Fisherman Above the Reef.* Both of these works are believed to have been done during the artist's second visit to Bermuda, the winter season of 1900-1901.

North Road, Bermuda (plate 58), was initialed and dated by the artist in 1900. This tranquil scene was quite possibly painted along North Shore Road in Smith's province near Flatt's Inlet, Flatt's Island and the parish of Flatt's. The small land portion visible on the right, just below the horizon, is probably Gibbet Island.

Interestingly, the subject originally called *Bermuda,* presently titled *Nassau* (plate 49), was described in William Howe Downes biography in this manner: "On the white beach in the immediate foreground are three rusty cannons; deep blue sea beyond. A sailboat, manned by two Negroes, is near the shore, and several other vessels are farther out. In the distance is a line of brown shore."[6] (It is reported that there were cannons along the beach on the south shore below Wreck Hill near Somerset Island, Bermuda.) The work bears the date of January 1, 1899—but could the artist have made an error, as happens quite often when a person writes a date beginning the new year.

Another typical titling problem is found in *Sloop, Nassau,* dated 1899 (plate 50), which for years carried the designation of having been painted in Bermuda. More recently, art historians have altered their thinking for various reasons. This subject and the above mentioned work are very similar in mood and setting.

This confusion as to titling and location brings to mind the request of Homer's brother Arthur that a list of his paintings be made and their whereabouts recorded. Downes relates the incident in this manner:

Asked if he knew where his works were, and who owned them, Homer replied that he knew where most of them were. 'Well, then,' said his brother (Arthur) 'why won't you make a list?' 'Why should I, so long as I know?' 'But nobody else will know after you are gone.' 'After I am dead I shan't care.'[7]

Beam had this to offer as to titling: "It is not surprising that with so many watercolors from his brush, many of them rather similar in theme and locale, some misnaming should have occurred. And particularly since he himself was occasionally guilty of mistitling, or of dating a picture when he finished and signed it, sometimes years after it had been started."[8]

This is, therefore, consistent with the conclusion which another biographer had reached that many of Homer's tropical pictures do not attempt to tell too much, but rather leave something to the imagination.

Following his return from Bermuda during his painting session of the winter of 1899-1900, Homer served with his friends John LaFarge, J. Alden Weir and seventeen other artists on a national arts jury for the selection of American works to be exhibited at the Universal Exposition again being held in Paris during the late spring of 1900.

He returned to Bermuda again the following winter and with the completion of his Nassau works, had more than forty watercolor paintings done "with the freedom and maturity of his new style."[9]

Prizing both series very highly, he was not eager to immediately expose them to exhibition. He worried that the public might again reject his bold tropicals of these later years as they had his earlier works.

He set a price of $4,000 on the

The only known photograph taken of Winslow Homer at work in his Prout's Neck studio. On his easel is the unfinsihed oil painting, *The Gulf Stream*, which was completed in 1899. It is believed the photo was taken during the winter months as the artist is warmly attired. Photo courtesy of the Homer Collection, Bowdoin College Museum of Art, Brunswick, Maine.

whole lot, with the hope of interesting a museum in their purchase. In 1901, Homer selected twenty one from this group and sent them to the Pan-American Exposition. They were awarded a gold medal there, but no buyer appeared. After they were returned to his Prout's Neck studio he wrote to his agent in New York:

That makes my winter's work of 1898 and 1899 complete. I shall leave them boxed as they are until such time as I see fit to put them out. The price will be $400 each!! for choice if I ever put them out again.[10]

He offered fourteen of the choicest to Knoedler's—*Flower Garden and Bungalow, Bermuda* (plate 55), was included, for exhibition at half the price quoted in his letter above. Some of the others he hung in Marr Cottage, the small building he said he'd erected to die in. In his Day Book he'd noted: "Murder comes now. To net me $200 each and all, good or bad, large or small."[11]

The reception that the oil *The Gulf Stream* (plate 14), received made him very angry. Homer felt that for a person "who was supposed to be the 'greatest American artist' " he was truly far from being understood.[12]

1. August L. Freundlich, "Winslow Homer's Sub-tropical America," *Exhibition Catalog Lowe Art Museum* (Coral Gables, Florida: University of Miami, 1968), p. 25. (Author's Note: This Bermuda subject was in the collection of the artist's brother and sister-in-law and was purchased in 1917, from Mrs. Charles S. Homer by Grenville H. Norcross. The watercolor in 1937, became the property of the Museum of Fine Arts, Boston, at the bequest of Mr. Norcross.).

2. Source: Bermuda, Department of Tourism & Trade Development. "The Bermuda Islands" (1972). (Author's Note: Recent photos of this area attest to the similarity to rock formations the artist painted over seventy years ago.).

3. Donelson F. Hoopes, *Winslow Homer Watercolors* (New York: Watson-Guptill Publications, 1969; third printing 1971), p. 52.

4. Author's Note: This subject could very well have been painted from Governor's Island as a ship was just entering St. George Channel with Paget Island in the background. [See also—Philip Conway Beam, *The Language of Art* (New York: Ronald Press, 1958), pp. 222, 363, and 795].

5. Author's Note: This seldom seen subject was sold in 1972 for $65,000 through The Kennedy Galleries. (See also—*Time* Magazine, November 20, 1972).

6. & 7. William Howe Downes, *The Life and Works of Winslow Homer* (New York and Boston: Houghton Mifflin Co., 1911), p. 260.

8. Philip C. Beam, *Winslow Homer at Prout's Neck* (Boston and Toronto: Little, Brown and Co., 1966), p. 215.

9. & 10. Jean Gould, *Winslow Homer A Portrait* (New York: Dodd, Mead & Co., 1962), p. 270.

11. Quoted from the artist's Day Book which is in the Homer Collection, Bowdoin College Museum of Art, Brunswick, Maine.

12. Gould, p. 273.

21

Return to Key West and Finding Homosassa: 1903-1904

Homer wrote a letter on either Saturday, December 5th or 12th, in 1903 to inform his younger brother of his plans for the winter season:

Dear Arthur

I decide to go direct to Key West. I have stateroom 20 upper deck "Sabine" go on board tonight leave early Sunday Morning—
I know the place quite well & its near the points in Florida that I wish to visit—I have an idea at present of doing some work but do not know how long that will last—at any rate I will once more have a good feed of goat flesh & smoke some good cigars and catch some red snappers.
I shall return through Florida & by May be at Scarboro.

Yours affy,
Winslow[1]

The ship Homer mentioned in this letter, the *Sabine*, was owned by the Mallory Steamship Company and sailed between New York, Key West and Galveston; the distance between New York and Key West is 1,150-miles and the running time was four days.

Homer probably arrived in Key West on the evening of Wednesday, December 9th or 16th, *or* the morning of Thursday, December 10th or 17th, depending upon when the harbor master would have assisted the captain in coming into the port. One definite record of time is found on *Stowing Sail* (plate 62), painted in Key West which the artist dated December 22, 1903. Having been to Key West previously, Homer probably had his lodging already arranged. The place was described in an impartial annual published in New York in 1884 as "one of the most interesting cities in the United States. Its winter climate is perfect; frost is unknown, even at night, and the tropical warmth of its days is delightfully tempered by steady trade winds. The city contains about 10,000 inhabitants, most of whom are engaged in the manufacture of cigars, sponging, fishing for the Havana market, or wrecking. It is located on an island of coral, of which the highest point is sixteen feet above the sea level. The luxuriant tropical foliage, overhanging and embowering the entire city, is a most novel and interesting sight to behold."[2]

According to Beam, Homer "must have gone directly afterward to Homosassa, Florida, a fisherman's paradise on the central west coast about 60-miles north of Tampa."[3]

The Key West subjects—to which Homer had alluded in his letter to Arthur with the sentence: *"I have an idea at present of doing some work but do not know how long that will last"*[4]—

Steamship *Leona* of the C. H. Mallory Company, later renamed the *Sabine.* This is how the vessel appeared during the Spanish American War as a troopship. Winslow Homer had cabin #20 on the upper deck during his trip to Key West in 1903. The photo is courtesy of the Mariners Museum, Newport News, Virginia.

were strictly marine in nature. They were the artist's continuation of a theme he began with one watercolor of 1885-'86, *Hauling Anchor* (plate 28).

Fishing Boats, Key West, dated 1903 (plate 63), also seems to bear the notation of "Dec." in the lower right corner.

Key West, Hauling Anchor, dated 1903, also known as *Weighing Anchor* is another of his paintings done in the final weeks of that year.[5]

Schooners at Anchor, Key West bears the designation "Key West, Dec. 1903" in the lower left corner.[6] *Taking on Wet Provisions* with the schooner's stern bearing the name, Newport, Key West, is another of Homer's 1903 watercolors.[7] One subject titled simply *Key West* (plate 61), is not dated but has every indication of having been a part of this particular period. *Fishing Boats, Key West,* dated 1904 (plate 64), apparently was done early in January and completed his marine studies at the Old Island city.

The next letter Arthur received from his brother came from Homosassa and was probably written immediately after his arrival.

Dear Arthur

Delightful climate here about as cool as our September—fishing the best in America as far as I can find

"Cavalle"
Yellow tail silver body.
Channel Bass
Look like a new $20 gold piece.
Trout
Black Bass
Sheepshead
I shall fish until the 20th then my guide has another engagement and I shall take my own boat and work half the time and fish on my own hook. I have not done any business this fall so far—& I shall only paint to see if I am in up to it—& with a chance of paying expenses—I have not heard a word from Charlie yet—I suppose you are all well—Do not send me any kind of newspapers—I have all I want.

Yrs. affy,
Winslow[8]

Homer probably traveled from Key West to Homosassa by boat via Cedar Key and then took a train on the Florida Railway and Navigation Company's line from Cedar Key down to Homosassa.

There were also stage lines connecting at Brooksville and Ocala. The spot was considered unsurpassed as a winter resort.

A book published in Baltimore in 1885, had this to report about Homosassa:

Weighing Anchor, is another of Homer's Key West subjects of 1903. Offering livelier tones of blue, it portrays a two-masted craft with her sails flapping gently in the trade winds while the anchor is being hoisted. Privately owned, this black and white photograph is courtesy of Bradley Olman.

The gulf coast of Hernando (county) has long been celebrated for its comparative healthfulness, and it is a familiar sight to find families from the interior camping for the summer on the banks and islands of the Homosassa. (These were, in many cases, Indians native to the clay-lands further east of the river.)
The exceptional attractions of this region have induced a number of gentlemen, representing all sections of the country, to unite their efforts to open these rich resources to their appropriate uses. They have purchased some ten to twelve thousand acres of choice and selected lands along the whole course of the Homosassa River, including all the Yulee Estate, and they are now expending large sums in improving these natural advantages, so as to make them availale for visitors and settlers.[9]

The above mentioned number of gentlemen included one or two persons who were particularly responsible for Homer's having come to Homosassa in the first place. The Homosassa Inn where the artist lodged was the property of The Homosassa Company, an incorporated body of Floridians and New Englanders. The Company's president was General Joshua Lawrence Chamberlain, who during the Civil War had served as a colonel of the 20th Maine Infantry and had been involved apparently with the final campaigns of the Army of the Potomac, to which Winslow Homer had been assigned as artist-correspondent.[10]

Chamberlain later served as governor of the state of Maine (1866-1871) and was appointed president of Bowdoin College, Brunswick, Maine (1871-1883). Chamberlain, quite possibly, was in close contact with the Homer family at Prout's Neck. (From 1884-1889 Chamberlain was deeply involved with railways and other enterprises in Florida.)

Another resident from the Neck, Frank Coolbroth, according to Beam, was "an amateur naturalist—knew Florida, for he went there every winter on business;"[11] he might also have been an integral part of the Homosassa Company, serving as one of its governors. Another New Englander, John C. Holman of Boston, was secretary for the company.

J.R. Campbell, of the St. James, Jacksonville, was a vice president and Col. John F. Dunn, of the Bank of Ocala, served as treasurer. (Dunn also acted as agent for the organization and the name Dunnellon given to the community in Marion County near Rainbow Springs is a combination of his last name and his wife's first name, Ellon.)

The Homosassa Inn was a single story frame building, built on a point of land called Shave's Landing jutting out in the river. Though not exceptionally spacious or as grandiose as it had been planned, since three shiploads of solid mahogany intended for its construction had been lost at sea during a summer storm, the Inn was a most congenial place for Homer to stay. (The artist must have enjoyed the place since there is record of his having returned to the Inn on two other occasions.)

The area was a great one for fishing, with two carloads of fresh fish being shipped to New York each week. Homer's mention of fishing with a guide until the 20th and his guide having then another engagement may have been due to the opening of the mullet fishing season, which was closed from November 20th until that date in January.

From Homer's mention to Arthur, of their brother Charlie, it could be that Charlie may have joined the artist for a portion of the winter in 1904. It is a well known fact that they often went on trips together and from the comments found in letters, there is a feeling that distance seldom prevented them from being together. Both Arthur and Charlie enjoyed Winslow's company and planned trips to outlandish areas where the fishing and/or hunting abounded just to provide him with the subject matter his craft continually required.

By the time Homer visited Homosassa, the population was approximately two hundred inhabitants and by 1910 there were three hundred-fifty residents, plus three general stores, one marine-store dealer, and the Williams Fish Co.

Homer found a wealth of subjects along the thousands of inlets and coves of the Homosassa River; *Channel Bass,* dated 1904, is one of the most famous of these. This watercolor depicts a hooked fish shooting through the turquoise water over the river's bottle littered bottom.

Palm Trees, Florida, dated 1904, is an impressive vertical watercolor close-up of nearly a dozen trees clustered together at the river's edge, one tree's trunk is bent and grows parallel with the water's surface; no human form appears in this work.[12]

Homosassa Jungle in Florida (plate 65), was given to the Fogg Museum of Art, Harvard University, by Mattie Homer in 1935 in memory of her husband Charles S. Homer, and his brother, Winslow. It is quite possible the fisherman tugging on the line was Mrs. Homer's husband and the other form that of a guide.

Homer was in an exuberant mood when he found the fishing excellent, and the weight of his almost seventy years could not keep him from fishing, wherever he found himself. . . . As to his painting, Beam concludes, "he was apparently up to it, for out of the trip of 1904 to Homosassa came some of his finest watercolors."[13]

Returning to Maine, through Florida, as he had written, Homer may have taken the Silver Springs, Ocala and Gulf Railway, advancing by way of Blue Spring and Crystal River. This would allow us a more logical explanation for the watercolor titled *Blue Spring, Florida* than previously mentioned. It is difficult, however, to give an accurate assessment as to the place and period to which it belongs.

1. Original correspondence in the Homer Collection, Bowdoin College Museum of Art, Brunswick, Maine.

2. C.K. Munroe, *The Florida Annual Impartial and Unsectional, 1884* (New York: 1883), p. 87.

3. Philip C. Beam, *Winslow Homer at Prout's Neck* (Boston and Toronto: Little, Brown and Co., 1966), p. 227.

4. William Howe Downes, *The Life and Works of Winslow Homer* (New York and Boston: Houghton Mifflin Co., 1911), p. 192.

5. Author's Note: Properly titled, *Weighing Anchor*, this privately owned work is an excellent example of Homer's profound simplification.

6. Author's Note: This watercolor is owned by the Dietrich Corporation.

7. Donelson F. Hoopes, *Winslow Homer Watercolors* (New York: Watson-Guptill Publications, 1969; Third Printing 1971), p. 78. (Author's Note: Owned by The Metropolitan Museum of Art, this marine watercolor offers a bit of old-fashioned good humor for the "wet provisions" are barrels of rum).

8. Original correspondence in the Homer Collection, Bowdoin College Museum of Art, Brunswick, Maine. (Author's Note: The original letter is enlivened with five crude but accurate sketches of the fish mentioned - a style Homer engaged in frequently in his letters).

9. M.B. Hillyard, *The New South, A Book on the Resources, Attractions and Development of the Southern States* (Baltimore: Record Printing House, 1885), p. 11.

10. Author's Note: My deep appreciation to the Maine State Library for affording data on Joshua L. Chamberlain and his memoirs.

11. Beam, p. 158.

12. Author's Note: This work is in the collection of the Museum of Fine Arts, Boston.

13. Beam, p. 229.

22

"A Kinder Man Never Walked the Earth . . ."

From late 1900, throughout 1901 and 1902, in brief but frequent letters to his agent M. O'Brien & Son in Chicago, Homer conveyed information concerning certain works either in progress or completed and in transit to them. Most of these subjects were his epics of the sea depicting the fury of the storms which hit so violently along the coast of Maine.

The relationship with the O'Brien gallery, which eventually proved to be a long and pleasant one, began shortly after the Chicago, Columbian Exposition of 1893 when they offered to represent Homer in the midwestern section of America. He replied to them on October 23, 1893, from Prout's Neck that he "would be glad to accept the offer when the time seemed opportune."[1] A later communiqué to them is dated October 19, 1900:

MESSRS. M. O'Brien & Son,

GENTLEMEN,—I have a very excellent painting, "On a Lee Shore" (here he gives a very slight pen-and-ink sketch of the subject), *39 × 39. The price is (with the frame) $2000, net. I will send it to you if you desire to see it. Good things are scarce. Frame not ordered yet, but I can send it by the time McKinley is elected.*

Yours respectfully,
Winslow Homer[2]

The painting, a wonderful representation of the Atlantic in its stormy mood as seen from the shore at Prout's Neck, was bought by Dr. F.W. Gunsaulus, president of the Armour Institute of Technology, who later loaned it to the Rhode Island School of Design for an exhibition in 1901. That institution purchased the marine masterpiece with the Jesse Metcalf Fund, an annual purchase fund for American works of art.

Upon realizing that a considerable demand for his pictures existed in Chicago, Homer decided to do a special commission for O'Brien & Son; the following letter relates to this undertaking:

Dec. 20th, 1900

M. O'Brien & Son,

GENTLEMEN,—I am extremely obliged to you for your kind letter, and the picture that you refer to I promise to send to you when finished.
I will look upon it in future as your particular picture. I do not think I can finish it before I have a crack at it out of doors in the spring. I do not like to rely on my study that I have used up to date.
But here is something that I can do. I shall have in about three weeks' time as many as six pictures all framed and on sale and exhibition. I will ship some of them to you, as the present holder should get sick

of them after two or three weeks' trial of sales. I show three at the Union League Club on Jan. 10th. I will let you have something this winter. I will notify you when I leave here.

Yours very truly,
Winslow Homer.[3]

The above letter was probably written just prior to the artist's departure for Bermuda via Boston and New York.

Approximately one year later, Homer wrote this to his Chicago dealer:

Scarboro, Me.
Dec. 3, 1901

My last letter referred to three photographs that were sent to me by the owner to be signed ("in very black ink," etc.).
Anything written or printed under a print or picture takes the attention from it, and if it is very black or white in any marked degree will utterly destroy its beauty.
When I received these photographs, I found much to my disapproval that a photographer had put his name and imprint immediately under the right hand side of the print (the place for the artist's signature), in a most pronounced manner. (Pen-and-ink sketch here.) *I have forgotten his name, but he is not the man who took the negative.*
The place for the man's name, if he has any right to show it on an unpublished print, is here: (Pen-and-ink sketch showing the name at the lower right-hand corner of the mount.) *That incident is closed. It is about time that I received my picture the "Gulf Stream" back from Venice, and the beautiful frame on it will go on the O'B. Partic' picture directly I can get hold of it and finish the picture.*

Yours respectfully and very truly,
Winslow Homer.[4]

Hastily sketched underneath his signature was a lighted lamp with the notation, *6:30 A.M. Dec. 3.* Always an early riser, this was to let O'Brien know they were fresh on his mind, even before daylight.

Further correspondence explains the delay in Homer's completion of the "O'B. gem" as he began to refer to the subjects:

March 15, 1902

I have ordered M. Knoedler & Co. to send to you the two oil paintings, "The Gulf Stream" and "High Cliff," the one just home from Venice, the other from Pittsburgh. You appear to expect three pictures. These two are the only ones. I mentioned that the frame on the large picture would fit the "O'Brien"—but the O'B. is not finished. It will please you to know that, after waiting a full year,

looking out every day for it (when I have been here), on the 24th of Feb'y, my birthday, I got the light and the sea that I wanted; but as it was very cold I had to paint out of my window, and I was a little too far away,—and although making a beautiful thing—(here is inserted a rude sketch of a trumpet, marked "own trumpet")*—it is not good enough yet, and I must have another painting from nature on it.*
The net price to me on "High Cliff" is $2000 (two thousand dollars).
The net price to me on "The Gulf Stream" is $3000 (three thousand dollars).

Yours truly,
Winslow Homer.[5]

Fifteen days later Homer wrote:

I am in receipt of your letter. In reply I will say that I think it is quite possible that the O'B. picture will be the last thing of importance that I shall paint. The present "High Cliff" that you have is the best of the two or three oil paintings that I now own. I have many watercolors—"Two Winters in the West Indies," as good work, with the exception of one or two etchings, as I ever did.
With the duckets that I now have safe, I think I will retire at 66 years of age, praise God in good health. I take note of your flattering request for photographs of myself. I think I may have one made, and I will send it to you.

Yours very truly,
Winslow Homer.[6]

That summer season Homer went up into Canada again. Some of his efforts during that time were *Shooting the Rapids,* a watercolor in the Brooklyn Museum's collection, and *Saguenay River, Grand Discharge, St. Johns, P.Q.,* a work which plays "around the theme of swiftly moving water. This is no easy task, and requires a nice balance between order and turbulence to avoid making the water look chaotic on the one hand or frozen on the other."[7]

In Canada, Homer faced "the same painting problems posed by the deep Adirondack woods and the tropical jungle—communicating the character of the woods through synthesis and suppression of details, avoiding the clutter in which one 'could not see the forest for the trees.' "[8]

It was September 27th, 1902 before he wrote again to O'Brien:

GENTLEMEN,

I find on looking up my drawing, that I have not seen for fifteen years, that I have only twelve. These I have sent to you today by the American Express. Please handle them very carefully until they are framed with a narrow half-inch black wood

frame and in the same mats in which they are; in fact, open them once, and take the measure, and then put them away until the frames, (are ready) and after that show them together *as you see fit. (They should be) sold to some Western museum.*
As quick sketches from nature (untouched)—you cannot beat them.
(Here a pen-and-ink sketch of a man blowing his own trumpet vigorously.)
I will take $400 net for the lot.

Yours very truly,
Winslow Homer.

P.S. Will you please acknowledge receipt of these drawings when you receive them? [9]

We can assume the dozen paintings Homer mentioned in his letter were some of his early Bahamian watercolors of 1884-'85 which were in part reproduced in black and white in *The Century* in 1887.

On October 29th, he wrote that he worked on the "O'B." from 6 to 8 o'clock that morning—that *the long looked for day arrived.* He finished the painting—making that the fourth two hour painting-from-nature session on the canvas.

This is the best picture of the sea that I have painted, he proudly wrote to his agent. *The price that you will charge is five thousand dollars—$5000. The price net to me will be $4000."* And, then the artist adds, *This may be the last as well as the best picture.* His reasoning: *I have rents enough to keep me out of the poorhouse.* Then, he continues, strictly businesslike: *Now all you have to do in reply to this is to notify me when you get the two pictures back from Des Moines, and I will then tell you what to do with them, and send the "O'B." picture.* [10]

His next letters of November 6th and 14th then produce thorough instructions concerning the handling of the forementioned pictures and the framing of the "O'B."

GENTLEMEN,

I am in receipt of your favor of Nov. 3rd. Please take out of its frame "The Gulf Stream;" pack it in a strong case, not more than three inches deep, with a cover put on with screws. Ship it to me at (Rubber stamp giving his name and address at Scarboro, Maine.) *Directly I receive it, I will put into the case the O'B. Gem, and ship it to you.*
In the meantime you will please put the frame in good order. The two pictures are the same size.
If I find there is any difference in the size of the two canvases, I will telegraph you, and have the frame whittled to suit, so there will be no delay in putting the canvas in the frame as it is safer there.
When I send the picture I will give you the few wishes

I have in the matter of the exhibition. It will only concern its protection from being used by others before it is widely shown.
I wish it sent to the Union League Club, New York, under your protection, for the loan exhibition of American artists. I will get an invitation for that purpose.

His next communiqué begins:

The O'B. leaves here by the American Express at 3 P.M. If it is damp when you receive (it) and the canvas wobbles, do not key it up, as the keys are glued in to the stretcher, and everything is in perfect order. Just put it in a warm room.
There was a sleet storm yesterday, but beautiful today, so I start O'B., and glad to get it out of my sight before I finish it too highly and spoil it. I hope the original member of your firm is still alive, after all these tedious years of waiting, and that he will be on hand to greet the O'B.

Yours truly,
Winslow Homer

Through these letters[11] we have had a glimpse of the aesthetic and practical conscientiousness of this artist. It demonstrates the orderliness of his life which he carefully maintained year after year.

Homer fought desperately throughout his career not to be blinded by prejudice—that of his chief critics, his own or other people's—for he felt this might have been transferred to his own work. He was occasionally affected by a sharply critical review and wrote terse letters, laying down certain precepts which he considered sacred.

To a dealer who passed on to him some of the comments concerning the horror of his oil subject, Homer wrote:

The criticisms of "The Gulf Stream" by old women and others are noted. You may inform these people that the Negro did not starve to death. He was not eaten by the sharks. The waterspout did not hit him. And he was rescued by a passing ship which is not shown in the picture.

Yours truly,
Winslow Homer.[12]

Most of the critics of his day wanted pictures which told a story, and the more story the better. Homer's subject seemed vulgar by Victorian standards and lacked the polite elegance of the day.[13] Criticized as he was for *The Gulf Stream* (plate 14), there were many who, despite harsh words in print, considered Winslow Homer to be the strongest American artist of his period.

During the last ten years of his life, he began to receive the prestigious acclaim and timely recognition that he'd long felt he was entitled. It still riled him inwardly to recall the public's misunderstanding of some of his subjects.

After remaining unsold for seven years, *The Gulf Stream* appeared at the National Academy. Homer wrote to Alden Weir that he'd kept away from the Academy's exhibitions "by the fear of the corridor and the impropriety of my trying to make terms as to placement of my work,"[14] he hoped now that was not necessary. Just shortly after the exhibit was opened, word was sent to him that the Metropolitan Museum of Art had been advised by the entire jury panel that the oil be purchased for its permanent collection. His most controversial work now had a home. Homer was indeed elated![15]

Also in 1906 three other of Homer's tropicals were on public view; *Channel Bass* (1904), *Black Bass, Florida* (1904) and *View from Prospect Hill, Bermuda* (1900) were included in the Pennsylvania Academy of Fine Arts' watercolor exhibition that spring.

Homer's works were being greatly admired even though one critic found the "absence of formulas" in his art baffling. Henry James was another who classified the artist as a man with "no imagination, but contrives to elevate this rather blighting negative into a blooming and honourable positive . . . Mr. Homer has the great merit, moreover, that he naturally sees everything at one with its envelope of light and air."[16]

Throughout his lifetime, Homer remained consistently true to the commitment of an artist to nature and during a visit to Prout's Neck, John Beatty, director of the Carnegie Institute quoted him as saying:

When I have selected the thing carefully, I paint it exactly as it appears—You must wait and wait patiently, until the exceptional, the wonderful effect or aspect comes. Then, if you have sense enough to see it—well, that's all there is to that.[17]

During the spring of 1908 a special group of Homer's works appeared as a "unique feature" of the 12th Annual Exhibition at Carnegie Institute in Pittsburgh; included among the twenty two oils were two of his tropical subjects: *The Gulf Stream,* dated 1899 (plate 14) and *Searchlight, Harbor Entrance, Santiago de Cuba* (plates 19 and 20).[18] This was probably the most important one-man show ever held during his lifetime—half of the paintings were lent by private collectors and the other half came from leading museums throughout the eastern and midwestern states.

On March 7th, 1908, Homer wrote the following letter to his younger brother:

Windsor Hotel
Jacksonville, Fla.

Dear Arthur

I tried to get bro Charlie on Telegraph as I had not heard from him for a month and as I failed to get any answer in two days I sent one to you. I have now recd one from West Townsend & glad to know that he is in this country.
I sail from here soon for New York.
As this is the time when people who skin summer boarders both on cottages & grub are hard up—I take this opportunity of making you a present of the enclosed check—

Affectionately,
Winslow Homer[19]

Presumably, while in Florida, the artist visited Enterprise or Homosassa, but apparently did no painting—preferring to fish instead.

Homer spent the first three months of 1908 in Florida and when he returned to Prout's Neck at the end of March found that his studio had been burglarized—the place was completely ransacked. Paintings in the loft were left untouched, their value undoubtedly not realized by the intruders, but a watch his mother had given him years earlier was taken. This loss was particularly upsetting to the artist.

In May, he awakened one morning unable to see well or to tie the usual sailor's knot in his tie; he finally managed to dress himself and walk gingerly over to Arthur's nearby cottage. Homer could not imagine what was the matter. He had suffered a slightly paralytic stroke but did not know it. Arthur and Alice, deeply concerned, managed to persuade him to stay with them for the time being. He allowed them to wait on him for two weeks while he rested and regained a little of his former strength. His eyesight was affected and he joked to Charlie of "having one eye on the pot and the other on the chimney."[20]

That summer the Homer brothers went to the Adirondacks and by fall it was evident that Homer's illness had not permanently affected his keen vision or his firm hand. Everything was apparently normal and he was quite thankful. If he did notice an occasional twinge in his stomach he never mentioned it to the family.

From November 23rd to December 20th, 1908, four of his paintings owned by Dr. and Mrs. George Woodward of Philadelphia were shown at the Pennsylvania Academy of Fine Arts' sixth annual watercolor exhibition. These were two works bearing the same title of *Prout's Neck, Maine,* and also *Spanish Flag, Santiago de Cuba* (plate 18) and *Volante.*[21] At the time of this exhibition, Homer was still at Prout's Neck and wrote to his older brother from there on December 8th.

Dear Charlie,
I do not think I shall leave here before January then I shall go directly south to Homosassa after about three days in New York.
I am painting when it is light enough—on a most surprising picture but the days are short & sometimes very dark—I am very well—but how are you & Mattie?
Affectionately,
Winslow[22]

Quite Japanese in style, this most surprising picture is a large oil study whose subject depicts a brace of wild ducks Homer had received from his friend Phineas W. Sprague, who probably intended them for the artist's Thanksgiving dinner.

Homer found "their plumage so handsome he was tempted to make a painting of them." He hired neighbor William Googins to "row off shore in a boat and fire blank charges up toward the cliff where he stood watching."[23] Homer wanted to accurately portray the fowl as they would appear when hit over water, shot in quick succession by a double-barreled shotgun—thus the title, *Right and Left.*[24] He finished his striking, richly styled painting on January 7th, 1909 and took it with him to New York where his dealer M. Knoedler sold it almost immediately to a private Philadelphia collector.

His face had begun to show distinct signs of aging; the internal pains in his abdomen were perhaps becoming increasingly troublesome. William Howe Downes was after Homer for details of his life for a biography Downes had been working on for a number of years. Homer stalled the man off as long as he could and eventually wrote him on August 13th, 1910, the following:

No doubt, as you say, a man is known by his works. This I have heard at many a funeral. And no doubt in your thoughts (it) occurred to you in thinking of me. Others are thinking the same thing. One is the Mutual Life Insurance Co., in which I have an annuity. But I will beat you both. I have all your letters, and will answer all your questions in time, if you live long enough.
In reply to your recent letter I will say that I was in Tynemouth in 1881 and 1882, and worked there.
Yours very truly,
Winslow Homer.[25]

The artist's defiant prophesy "I

will beat you both," sadly did not come about and Downes never received the answers to his questions.

An ill man throughout the summer of 1910, Homer evidently endured his stomach pain in stoic silence. When it became fully evident to his brothers and their wives that he was suffering mental and physical anguish, they made plans to make him more comfortable and provide him better care at Arthur's cottage. Homer refused.

Relentlessly stubborn he refused also to have a woman wait on him—so the family obtained a male nurse from Boston to care for him at his studio. Dr. B.F. Wentworth examined the artist and found him nearly exhausted from a loss of blood, which on closer examination proved to be caused from the rupture of a blood vessel in his stomach.

Winslow had the will to live and he resisted this final illness as long as he was able. The first hemorrhaging was checked for a time; a second followed during which he lost his eyesight and was delirious for several days. Then, his mind cleared but his sight did not return. Gradually, he regained some strength and began to make plans for the coming winter; perhaps Charlie and he could go to Homosassa for some bass fishing. The days were probably filled reminiscing and joking with Arthur and Charlie, recalling their old pranks and childhood experiences at Cambridge, speaking of the things they had not talked about for years.

Up to the morning of September 29th, there was a "real or apparent improvement." Night and day, Homer was attended by excellent nurses and all of his dearest relatives were at his side. Then, suddenly an alarming change took place. The end was evidently near. He expired at 1:30 p.m. that day at the age of 74. He had passed away literally "in the harness" for he died in his own studio, "a most fitting place—like a soldier dying on the field of battle, with the flag waving over him, a glorious passing of the brave, indomitable spirit."[26]

The Associated Press correspondent at Portland sent out a laconic dispatch informing the nation that Homer was gone. From Atlantic to Pacific the news was received with "unfeigned and universal sorrow."[27] In the hundreds of press tributes that followed, the writers took note of the American quality of his art.

There was a small, brief funeral and following this the artist's ashes were placed alongside his mother's grave in the southeast corner of the family lot at Mount Auburn Cemetery in Cambridge. The plot overlooks Consecration Dell on Lily Path with a low red granite boundary enclosing the area. It was fitting indeed that the remains of Winslow Homer should be returned to

Cambridge, for this was his boyhood home where he had learned to love and respect nature.

In the Boston *American,* a Portland correspondent reported on October 1st that "real tears" would be "shed among the fisherfolk of the little village of Prout's Neck, for they loved him as a brother."[28]

The artist's estate was initially estimated by his older brother Charlie as being in the neighborhood of $40,000. Had Homer not given away so much money to those in need among his fellow Prout's Neck residents, the estate probably would have been considerably larger. Frank Coolbroth, Homer's old friend and frequent walking companion, later appraised the estate at about $70,000. He left the world a wealth of art that even today continues to grow in esteem and value.

In the winter of 1911, Boston and New York, the two cities most closely related to Homer's art career, "made haste to do him honor," according to Downes. Special exhibitions of his works took place simultaneously at the East Gallery of Boston's Museum of Fine Arts and at the Metropolitan Museum of Art's Gallery XX in New York. The latter exhibition was a collection of fifty-two works (twenty-eight of which were watercolors considered representative of the artist's achievements in various periods and phases of his art). Wisely, "no attempt to make the collection, complete, exhaustive"[29] was offered. Eleven of the works included in the Memorial Exhibition by the Metropolitan, later acquired for its permanent collection, were tropicals.

Many beautiful recollections appeared in periodicals around the world and one of these was written by well-known critic, C. Lewis Hind: "It is as a master that I always regard Winslow Homer. . . In daily companionship with the ocean he led a solitary life."[30]

1. Philip C. Beam, *Winslow Homer at Prout's Neck* (Boston and Toronto: Little, Brown and Co., 1966), p. 116.

2-6. William Howe Downes, *The Life and Works of Winslow Homer* (New York and Boston: Houghton Mifflin Co., 1911), pp. 209-216.

7. Beam, p. 251.

8. Ibid., pp. 127-130.

9.-11. Quoted in Downes, pp. 216-219.

12. Quoted in Beam, p. 172.

13. Albert Ten Eyck Gardner, *Winslow Homer American Artist: His World and His Work* (New York: Clarkson N. Potter, Bramhall House, 1961), pp. 193-196.

14. Quoted in Lloyd Goodrich, *Winslow Homer* (New York: George Braziller, Inc., 1959), pp. 53-54.

15. Downes, p. 259.

16. & 17. Quoted in Jean Gould, *Winslow Homer A Portrait* (New York: Dodd, Mead & Co., 1962), p. 283.

18. Author's Note: This superb oil is owned by The Metropolitan Museum of Art, a gift of George A. Hern. (Plates 19 and 20) are the preliminary efforts to the oil which was painted over an extended period of time and signed, Homer 1901.

19. Original correspondence in the Homer Collection, Bowdoin College Museum of Art, Brunswick, Maine.

20. Quoted in Gould, p. 289. (Author's Note: The artist further expressed that this situation was "a new departure in the world").

21. Author's Note: *Volanté* was described by Downes as being a "quaint" Cuban cab.

22. Original correspondence in the Homer Collection, Bowdoin College Museum of Art, Brunswick, Maine.

23. Gould, p. 289. (See also—Beam, p. 249).

24. Author's Note: This painting is now in the National Gallery of Art, Washington, D.C.

25. Downes, p. 249. (Author's Note: The art critic's final letter from Homer).

26.-30. Downes, pp. 252-257.

Plates

Plate 1
SCHOONER AT SUNSET (1880)
Watercolor 13¾ × 9¾″
Signed: LR; HOMER 1880
Fogg Art Museum, Harvard University
Grenville L. Winthrop Bequest

Winslow Homer loved sunsets. It has been said he would walk backwards across Saco Beach near Prout's Neck so as not to miss any nuance of the visual drama enfolding, and when recording the relationship of clouds and light, he wanted them exactly as they occurred, believing that nature knew best how to arrange such a scene.

This work demonstrates effectively the range of color and treatment Homer was attempting to achieve prior to his visits to the tropics. The subject also gives us an insight into the artist's experimentation with the dry brush technique of watercolor, skipping over sections to leave some of the paper exposed. Its unfinished look, however, brought adverse comments from the critics.

Painted in 1880, four years before his first trip to the Bahamas, *Schooner at Sunset* is not one of Homer's more popular pictures. It is one of nearly 30 watercolors he painted at Gloucester during his two visits there in 1873 and 1880. Other paintings in similar themes completed at approximately the same time as this watercolor are *Gloucester Harbor, Fishing Fleet, Boys Beaching the Boat, The Green Dory* and *Evening on the Beach.*

Plate 2
PATH TO THE BEACH (1885)
Watercolor 20 × 14″
Courtesy of Wildenstein & Company, New York.

This little known painting reflects a delightfully placid mood. Possibly Homer was finding release from tension; getting away from the task of serving on jury duty, from the nearly overwhelming grief of the loss of his mother, Henrietta, or from the wintry winds of Maine's rocky coast.

Though considered unfinished this watercolor, in its precise yet hazy delineation of the palm trees' fronds and the penciling over of the cactus in the right foregcreates for viewers an almost ethereal vision. It was painted at Nassau on his first voyage there.

Note: *A Road in Nassau,* another early Bahamian work, presents a more pronounced impression with "a striking pattern of old white walls gleaming against emerald foliage and cool distances in a perfectly peaceful mood," according to biographer Philip C. Beam. This was the type of subject made-to-order for Homer—exercises in simplication and elimination which he repeated time after time.

Plate 3
HEMP (1885)
Watercolor 19¼ × 13½″
Signed: LR; Winslow Homer 1885
Courtesy of Wildenstein & Company, New York.

This remarkably effective painting is from Homer's first visit to Nassau. The figures, greatly simplified, are blocked in. The gray sky and the sand in the foreground are rendered in a flat wash. Tall shafts rising from the cacti and the plants' spiky leaves are done with precise outlines. The composition is interesting. The line of bushes is broken by the spear-like flower shafts rising above and by the diagonal line formed by the walk-way on the left. Homer's handling is, however, somewhat stiff and labored. He was still too preoccupied with details to allow his brush great freedom, and to simply summarize the whole with a few strokes.

The Century Magazine, February, 1887, carried a small black and white reproduction of this watercolor study with the title *Growing Hemp.* Accompanying it were eight other of Homer's Bahamian paintings of 1885.

Winslow Homer

Plate 4
TORNADO, BAHAMAS (1885)
Watercolor $19\frac{3}{4} \times 13\frac{1}{2}''$
Signed: LR; Winslow Homer
Collection of Mortimer Spiller

Recognizing the tragic and violent side of life and nature in those balmy climes, Homer painted this watercolor during his initial voyage to New Providence.

This work appeared also in black and white in *The Century Magazine* as an illustration for an article touting the area's potential for winter vacationers. It must be realized that these early Nassau watercolors presented viewers a comprehensive idea of the life and landscape of the Bahamas, a part of the world most alluring from the point of view of paintable materials, which had not before been exploited.

The black and white title of this work, as it appeared February, 1887, was *A Hurricane.*

Plate 5
ON THE WAY TO THE MARKET,
BAHAMAS (1885)
Watercolor 19½ × 13¼″
Signed: LR; Winslow Homer 1885
The Brooklyn Museum
Gift of Gunnar Maske in memory of Elizabeth
Treadway White Maske

Homer was fascinated by every facet of Bahamian life, from the bustling harbour front at New Providence Island to the narrow pathways of the Out Islands. This charming painting captures the graceful gait of a Negress enroute to the native outdoor market to sell a pair of chickens. Her travel takes her past a stone wall. Behind the wall grows a handsome coconut palm, blazing poinsettias, and fruit-filled orange and grapefruit trees. The artist's horticultural instincts allowed him to miss no small detail in presenting viewers this profusion of lush tropical growth.

A trip to Harbour Island during his 1884-1885 sojourn in the Bahamas brought forth another watercolor entitled *Native Woman Cooking,* which might be considered a companion to this subject. This second painting depicts a woman wearing an off-white skirt and brownish red shawl cooking before a pale blue outdoor oven. The foreground is gray with reddish brown and green accents. The sky is soft blue to the right and whitish gray at left. The artist's signature appears at lower left and on the reverse side is the inscription: Harbor (sic) Island, Bahamas, 1885. This was no doubt painted about the same time as *Glass Windows,* the painting of Eleuthera's narrow land ridge which formed the spine of the island.

The Museum of Fine Arts, Boston, obtained *Native Woman Cooking* in 1939 as a gift from Mrs. Robert B. Osgood.

HOMER
1885

Plate 6
SPANISH BAYONETS (1885)
Watercolor 20¼ × 15″
Signed: LL; HOMER 1885
Fogg Art Museum, Harvard University
Grenville L. Winthrop Bequest

Undoubtedly, this is the panorama which confronted Homer's keen eye while he was seated beside the shoreline of New Providence Island, within sight of Fort Charlotte. Just off center is Hog Island Light, a beacon built in the early 1800's to direct ships safely into Nassau Harbour.

The plants in the foreground apparently intrigued the artist. He was consistently aware of botanical and horticultural objects throughout his lifetime—an awareness stemming from his mother's deep, abiding interest in portraying plants and flowers.

In the tropics, Homer found many plants and trees completely dissimilar to anything he'd ever seen in Maine. The stalks of the Spanish Bayonet are exceedingly sharp and stiff—and are actually very dangerous! When Homer arrived at Nassau late in December, the bloom had already dried and withered. The delicate, waxy, candle-like flowers were long gone and, therefore, of no interest to the artist. They simply disappear out of sight, off the paper. The sky is grayed and conveys a slight chill but the waters do not seem to be ruffled by winds.

"Glass Windows"
Winslow Homer

Plate 7
GLASS WINDOWS, BAHAMAS (1885)
Watercolor 20 × 13 15/16″
Signed: LR; Winslow Homer
"Glass Windows" Bahamas
The Brooklyn Museum

The subject of this watercolor also appeared as a wood engraving in *The Century Magazine,* February, 1887, with the title *On Abaco Island.* The accompanying text states: "The ocean works it (the rock) into fantastic forms, of which we have an illustration in the famous Hole-in-the-wall on Abaco Island. This is an opening in the calcareous rock, through which the setting sun, blazing in its tropical majesty, at times produces a picture leaving an impression never to be effaced."

There is doubt now as to where this particular rock archway was located in the Bahamas. Homer gave it the title *Glass Windows* and according to Alison Grandfield in *The Bahamas: Island by Island,* such a name had been used by the natives for an arch on the eastern coast of Eleuthera. "Once a land bridge formed a perfect window in the rock which sailors could see through as they passed. Winslow Homer painted this natural phenomenon before it was washed away in a hurricane. Today a concrete bridge spans the gap."

Interestingly, too, this rock-formation watercolor by Homer is quite similar in concept, detail and feeling to *Arched Rock, Capri,* an 1848 study by Jasper F. Cropsey (1823-1900) who was one of the founders of the American Society of Painters in Water Colors. Homer also became a "founding member" in 1866, when the group received the more euphonious name of the American Water Color Society.

Plate 8
SHARK FISHING (1885)
Watercolor 20 × 13⅞″
Signed: LL; HOMER
Collection of Laurance S. Rockefeller

Continually "aware of the savagery latent in nature," Homer had seen the ferocious sharks lurking just beneath the calm waters off Nassau Bar. On his first visit to the Bahama islands he started his series of studies which depicted these creatures "in relation to the native fishermen." The idea was developed "only tentatively at first by depicting some natives capturing a shark . . ." Beam states.

In this painting, Homer depicted what he saw "as exactly, as forcefully and as handsomely as possible." The artist is conveying to his audience how men ventured out in their small boats and caught the frightful sharks which infested the harbour waters throughout the tropics. This challenge of capturing an instantaneous view of nature was ever fascinating to him.

A slight, but extremely energetic man, Homer loved the outdoors and found the tropics a peculiarly masculine world. To him a picture which didn't tell a story was as incongruous as a sentence that didn't contain a subject and a predicate.

HOMER 1885

Plate 9
THE CONCH DIVERS (1885)
Watercolor 20 × 14″
Signed: LL; HOMER 85
LR; HOMER 1885
The Minneapolis Institute of Art
The Dunwoody Fund

This fine watercolor gives the impression of a group of natives going about a daily task quite unaware of the artist. His three central figures appear to be in progression—almost demonstrating the movements of a single human form. The dark curved shapes standing, bent and hunkered over the gunwale, contrast sharply with the still sea and open sky. The picture is deeply satisfying, yet unassuming and the storytelling element is virtually eliminated.

In this painting, Homer concentrated on the visual pleasures of composition, atmosphere, and color. And, when studying the details of this watercolor, one realizes how much he loved painting the superbly developed Blacks of the Bahamas.

There also is another familiar mark of Homer in this painting—the double signature at lower left and on the gunwale on lower right.

Plate 10
SEA GARDEN, BAHAMAS (1885)
Watercolor $8\frac{7}{8} \times 15\frac{1}{4}$"
Signed: LR; HOMER 1885
Fogg Art Museum, Harvard University
Cambridge, Massachusetts

These two young Bahamians undoubtedly caught Homer's discerning eye as he strolled along the docks at Nassau Harbour. The artist had just arrived and everything around him was bright and new. The boy, probably just returning from a trip to a nearby coral reef, proudly displays his treasures of the sea to a girl aboard a single-masted craft beside his small skiff. The tilt of her face and slight smile attest to her admiration for the delicate gift he presents—a soft lavendar sea fan.

Pale in coloring and small in size, this watercolor was painted during Homer's initial trip to New Providence, the capital island of the Bahamas. The weather seems to be slightly overcast as is often the case when a northwester blows from off the warmer waters of the Gulf Stream. Along the sailboat's gunwales at the right edge of the painting lie stalks of sugar cane—well-loved by natives throughout the tropics then as now.

It is thought that no color reproduction of this painting has ever before been published. Yet, truly, it deserves careful consideration as an interesting study in human relations—a subject which Homer relished throughout his lifetime.

NASSAU

Plate 11
SPONGE FISHING/BAHAMAS (circa 1885)
circa 1885)
Watercolor 20 × 14″
Signed: LR; Winslow Homer
Canajoharie Library and Art Gallery
Canajoharie, N. Y.

All of his biographers have appraised Homer as a lifetime lover of boats. Beam emphasizes that he "admired the gleaming sloops of the tropics and made their trim lines a principal feature" in this sparkling painting. "His visual interest here was in the unifying effect of permeating luminosity and color. To stress these elements he reduced storytelling content and individual figures to minimum definition, and saturated the picture with color." Limpid and translucent watercolor washes exploit to maximum effect the white of the paper.

Created from an actual scene Homer had witnessed off New Providence Island, this masterpiece of radiance must have seemed an exaggeration to those who had never visited the tropics. Even the artist found this area to be a vast new world with intense warmth and nearly unbelievable colors. And, here he has conveyed the sunlight's unbelievable brilliance as it bounces off the waves.

Plate 12
STUDY FOR THE GULF STREAM (1885-1886)
Watercolor 10½ × 14½″
Cooper-Hewitt Museum of
Decorative Arts and Design
Smithsonian Institution/New York

When Homer was first visiting the Bahamas in 1884, he began his random studies of boats adrift. Later, he added fearsome sharks cruising hungrily around derelict sloops.

This watercolor depicts a derelict schooner and stalks of sugarcane lying on board. It is just slightly different from the boat appearing in *Derelict and Sharks, 1885* which shows the savage fish around the battered hulk.

Homer's fascination with sharks came directly from his viewing them first-hand while crossing the Gulf Stream and from sailing craft in the harbors of Nassau and Santiago de Cuba.

Plate 13
THE GULF STREAM (1889)
Watercolor $20\frac{1}{16} \times 11\frac{3}{8}$"
Signed: LR; Sketch W. H.
LL; 1889
The Art Institute of Chicago

This original watercolor is considered to be one of the few paintings of the tropics which Homer probably finished in his Prout's Neck studio. His shark-versus-man theme, perhaps unconventional, was thought to be a good deal too harsh for sensitive art lovers! As a storytelling picture, this work conveys stark reality as the artist saw and felt it. The viewer must simply draw his own conclusion.

It was not Homer's "nature to be circumspect from fear of public disapproval"—Beam asserts in a 1966 catalogue. He points out also that had this watercolor been exhibited instead of the later oil, (shown in the following color plate) "the shouting might have been less shrill." Neither painting, however, "avoids the gruesomeness of the topic, the watercolor does not underscore it."

Plate 14
THE GULF STREAM (1889)
Oil 49 × 28″
Signed: LL; HOMER 1899
The Metropolitan Museum of Art
Wolfe Fund

The theme Homer began in 1884 is more highly developed in this 1899 canvas. He worked for months until he felt the painting was ready to exhibit at Knoedler's late in 1900. The beautifully framed painting was then sent to Venice for the International Exhibition of 1901. In March, 1902, it was returned from Europe and sent by Knoedler and Company to M. O'Brien & Sons in Chicago.

Homer wrote to O'Brien's concerning the oil: "Why do you not try and sell the 'Gulf Stream' to the Layton Art Gallery, or some other public gallery? No one would expect to have it in a private house."

This oil is the only one of Homer's tropicals ever exhibited at the National Academy of Design and, following its 1906 exhibition there, was finally purchased by the Metropolitan Museum of Art for $4,500. According to Downes, the entire jury of the Academy recommended its purchase by the Metropolitan—a manifestation of the high esteem in which the artist was held by his professional brethren. One critic termed it "a rare thing . . . a great dramatic picture: Partly because of the horror it suggested without a trace of sentimentality, and partly because every object in the picture receives a sort of even, all-over emphasis that shows no favor to the dramatic passages; the story never overweights the artistic interest."

Plate 15
SANTIAGO DE CUBA (1885)
Watercolor
Signed: LR; Winslow Homer '85
West Point Museum
United States Military Academy

This subject, no doubt, shows how the harbour of Santiago appeared when Homer arrived from Nassau by ship at the end of February, 1885. In the background are the hazy, almost mystical outlines of the Sierra Maestra Mountains, a range that continues westward into the Caribbean and forms the three Cayman Islands, B.W.I. He wrote to his brother that the ancient city was "a red-hot place full of soldiers." He did a series of sketches and crayon drawings depicting the obsolete watchtower and cannon of Morro Castle. This was a crumbling citadel and as a bastion to protect the harbour it was no longer useful. His watercolors, as he gained control of his brush, conveyed the feeling of the antiquated fortress sedate in the hot West Indies sunlight.

A companion to this Cuban watercolor appeared in the 1936 memorial catalog for the Century Loan Exhibition sponsored by the Prouts Neck Association. Erroneously titled, *Town in Bermuda, 1885,* the painting was in the collection of Homer's nephew, Charles L. Homer. The year is accurate but the locale should be Santiago de Cuba. Another title was given some years ago, according to the Homer Collection at Bowdoin College; this was *Guanternode* and the painting was reported as being owned by Mr. & Mrs. Solton Engle.

Plate 16
VIEW OF STREET,
SANTIAGO DE CUBA (1885)
Watercolor 19⅝ × 10⅝″ (sight)
Signed: LL; Winslow Homer 1885
West Point Museum
United States Military Academy

Homer's Day Book carries a notation for Saturday, July 5, 1902, for this watercolor and refers to the title as being *Street in Santiago.* This was no doubt the same work which was included with *Spanish Club, Government Building, Diver, Nassau* and *The Road in Nassau* at the Boston Memorial Exhibition in February, 1911, and was specifically selected by the committee to show the beauty and romance of the tropics that Homer so ably depicted.

Santiago de Cuba, Street Scene (1885)—not the same painting as *Street Scene, Santiago de Cuba* 1885, (plate 17)—depicts the balconied, adobe buildings of the old city. Riding up a steep incline on horseback and pulling a burro behind is a male Cuban in the left foreground; far behind is the pale purple shadow of the mountains. Winslow apparently enjoyed depicting these picturesque perspectives of hilly Santiago.

Plate 17
STREET SCENE,
SANTIAGO DE CUBA (1885)
Watercolor $17\frac{1}{4} \times 11\frac{3}{4}$" (sheet)
Signed: LR; Winslow Homer 1885
Philadelphia Museum of Art
Given by Dr. & Mrs. George Woodward

Many of Homer's critics consider this typical of his street scene watercolors created at Santiago de Cuba. Enthralled by the old buildings, the unusual shades and rich textures of the stuccoed walls, the inviting reaches of the narrow streets, half washed in sun and half in cool shadow, Homer was truly in his natural element.

Interestingly, according to one account, the lady with the monkey on her shoulder as well as the woman carrying the parasol were not posed—they had simply entered the setting and Homer instantly recorded them with his deft brushstrokes.

HOMER
1885

Plate 18
SPANISH FLAG,
SANTIAGO DE CUBA (1885)
Watercolor 16¼ × 12¼″
Signed: LR; HOMER 1885
Philadelphia Museum of Art
Given by Dr. and Mrs. George Woodward

This colorful impression of Homer's winter visit to the ancient city on Cuba's southeastern coast has a somewhat unfinished quality since the lower portion of the señoritas' dresses are not visible through the ornate railing. Barely discernable, to the right of the shuttered opening, is a faint suggestion of a fourth figure. The artist brushed in the form very lightly—perhaps intending to complete the painting in his Prout's Neck studio.

This work is reminiscent of another balcony study by the Spanish artist, Goya, whose works Homer may have seen in Paris or London museums.

Note: *Spanish Flag, Cuba* was loaned, along with *Volante,* another Cuban subject, to the Sixth Annual Water Color Exhibition at the Pennsylvania Academy of Fine Arts by the Woodward's. Before Dr. Woodward died, he gave all but three of his Homer paintings to the Philadelphia Museum of Fine Arts—his three sons, Huston, George and Stanley received one painting each. *Custom House, Santiago de Cuba,* another of the 1885 series, is among those works retained by the family.

Plate 19
STUDY FOR SEARCHLIGHT, HARBOR ENTRANCE, SANTIAGO DE CUBA, (1885)
Pencil on paper 7⅞ × 4⅞″
Cooper-Hewitt Museum of Decorative Arts and Design
Smithsonian Institution, New York

Homer was intrigued with the shape of the round tower-like sentry's box projecting above the parapet at Morro Castle. The detailed, obsolete cannons extending horizontally across the composition's center offer a contrast to his rough sketch.

This was a solemn place; the artist was perhaps inclined to wistfully consider the stern, warlike old stronghold's more romantic days when Cuban political prisoners languished in its subterranean dungeons.

He retained this sketch and several others for a time and later completed an oil from them which was heralded by many of the leading art critics of the day as being among his best efforts.

Plate 20
SEARCHLIGHT, HARBOR ENTRANCE, SANTIAGO DE CUBA, (1885)
Pencil and chalk on paper, 18⅝ × 14″
Signed: LL; Drawn by Winslow Homer at Morro Castle Santiago de Cuba.

This study and the preceding sketch have been alternately titled *The Old Guns at Morro Castle, Santiago de Cuba.* Homer's final work on this subject was first seen at a loan exhibition held at the Union League Club, New York after the Spanish-American War of 1898 made Morro Castle famous.

A later oil painting, bearing the same title as this preliminary effort, was surmised to be an apparent comment on the Spanish-American War, with the distant searchlight representing the U.S. Fleet which destroyed the Spanish ships in that harbour. As in this pencil and white crayon sketch, the oil represents the sentry's box and cannons in dark shapes. Beyond them is a wide expanse of luminous, pale blue sky, "athwart which sweeps the wedge shaped light from the battery on the farther side of the channel, or possibly, from a

Plate 21
MORRO CASTLE (1885)
Watercolor
Signed: LL; Winslow Homer 1885
West Point Museum
United States Military Academy

This painting demonstrates Homer's tremendous capability for recording "the most delicate nuances as well as the most resonant tones, that he was in a singular degree endowed with the faculty of seeing justly and exactly the thing as it exists . . ." according to Philip C. Beam, curator of the Homer Collection, Bowdoin College.

One of Homer's earlier biographers, William Howe Downes, described Morro Castle as a "grim, medieval fortress . . . a sort of Caribbean Gibraltar, which frowns over the narrow entrance to the landlocked harbour."

Homer made a number of sketches of the old guns at Morro Castle and later, during the Spanish-American War, when Santiago Harbor became famous, he began work on the oil composition he titled *Searchlight, Harbor Entrance, Santiago de Cuba.*

This watercolor, along with thirteen other Cuban studies and fifteen Bahamian works, was included in Doll & Richards 1886 exhibition in Boston. It is rarely reproduced or seen by the general public.

Plate 22
THE GOVERNOR'S WIFE,
CUBA (1885)
Watercolor 19½ × 12⅛"
Signed: LR; Winslow Homer 1885
Museum of Art
Rhode Island School of Design, Providence

Until recently, this watercolor carried the location tagline: *Bahama Islands.* Research has uncovered not only the fact that the subject was painted at Santiago de Cuba but also the name and occasion for Homer to have painted the lovely model.

A curatorial assistant at the Rhode Island museum questioned the painting's locale and contacted this author as to what additional information might be available on the painting through my efforts. Contacts obtained via the Institute of Inter-American Studies at the University of Miami led to the son of the man who served as civil governor of the Province of Santiago de Cuba in 1885. Correspondence with this gentleman, now living in Madrid, Spain, brought about what can be considered positive identification.

Homer was the guest of General Luis Manuel de Pando y Sanchez and his beautiful young wife, the former Señorita Cecilia Armand Roch, a descendant of the french founders of the sugar mills Santa Cecilia and Romelie. The artist found an ideal setting along the palace's promenade, the pale blue forms of the Sierra Maestro Mountains forming a backdrop for the señora's dark-eyed beauty.

The de Pando's son wrote that his mother lived to the ripe, old age of ninety and was living in Madrid at the time of her death in 1954. Up until the Communist take-over in Cuba, the sugar mills remained in the control of the family.

Now, properly designated, *The Governor's Wife, Cuba,* is a wonderful example of Homer's ability to depict where and when he felt necessary even the most intricate patterning of lace and brocade, yet leave the remainder of the watercolor free and loose; focusing attention on the subject. The painting is another demonstration of how Homer celebrated youthful feminity.

Plate 23
CUBAN HILLSIDE (1885)
Pencil and Chinese white 22 × 16″ (frame)
Signed: LL; Winslow Homer 1885
The Cooper-Hewitt Museum of
Decorative Arts and Design
Smithsonian Institution, New York

On the island of Cuba Homer continued his studies of the tropical palms which he'd begun in Nassau. Sometimes he would show "the stately trees under storm conditions, when they sway so expressively with a kind of inner life," Beam commented.

In *Cuban Hillside,* he retreated to the surrounding countryside, away from the ancient cobblestone streets of Santiago de Cuba, and found a pastoral mountain setting.

His intense observation of palm trees becomes most apparent in some of his later finished watercolors like *Palm Tree, Nassau* (1899). Beam points out also that "no matter what part of the world he was in" Homer loved adding "the excitement and aliveness of high winds to his nature studies," and he was not long in trying his hand at the spectacular storms of the tropics.

"Royal Palms"

Plate 24
ROYAL PALMS, CUBA (1885)
Pencil & Crayon $17\frac{2}{5} \times 11\frac{3}{5}''$
Signed: LR; Winslow Homer "Royal Palms"
The Cooper-Hewitt Museum of
Decorative Arts and Design
Smithsonian Institution, New York

This quite oriental sketch and its companion, *Cuban Hillside,* are unusual in that neither work contains a horizontal: even the palm fronds high atop the vertical trunks are arranged in a series of diagonals.

In drawing *Royal Palms, Cuba* with its humid stillness, Homer moved in closer and gave more careful study—bringing the tree trunks and branches out in finer detail.

Note: *Coconut Palms, Key West,* a vertical watercolor, $13'' \times 16\frac{1}{2}''$, signed and dated 1886, was completed during Homer's visit to the Old Island city the next winter's season. This excellent work shows a handsome palm overburdened with near ripe fruit close-up in precise detail. It is an intensely serene study with no winds in evidence. Here, too, there are no strong horizontal lines; only a hint of a horizon is offered very lightly on the lower right side.

Plate 25
HILLSIDE WITH CLUMPS OF MAGUEY (1885)
Watercolor and pencil 9 × 14⅖″
The Cooper-Hewitt Museum of Decorative Arts and Design
Smithsonian Institution, New York

This is believed to have been sketched and painted in Cuba in the foothills of the Sierra Maestro Mountains outside Santiago. Almost Japanese in concept, it is interesting that Homer selected it. The plant was new to him, something he'd not found in his previous travels. The deep mauve and purple tones give the watercolor a somber mood. Homer, the artist, zeroed-in to bring the viewer a close look at these plants and yet maintained an aesthetic, almost abstract composition.

Plate 26
IN A FLORIDA JUNGLE (1885)
Watercolor 20 × 14″
Signed: LR; HOMER
Worcester Art Museum
Worcester, Massachusetts

Also known as *Florida Jungle* this is one of Homer's earliest efforts to depict "the tropical rain forest, that lush tropical growth that is so impressive it almost defied the efforts of writers and artists to do it justice," Beam assesses and goes on to state: "Like many another before and since his time, Homer tried rather unsuccessfully to paint it, and it was only after many attempts and much hard study that he gradually achieved results which satisfied him."

Nearly photographic in detailing, *In a Florida Jungle,* one of Homer's early visions of the mainland, captures the rich, verdant growth of the dense vegetation, but to Beam's feeling, "overdoes it." The truth is that, many portions of Florida's wilderness areas still remain exactly as Homer painted this watercolor and the denseness is accurately portrayed.

At approximately the same period, George Inness was coming to Florida and he painted the verdant growth with an ethereal, fantasy effect. Inness had spent many years studying in France at the Barbizon School and was undoubtedly affected by the influence of the popular painters there. Homer's short stay in France in 1867 brought about a slight change in his works, but it was more in the range of the effects of light and shadow as well as the design influence from the Japanese prints at the Universal Exposition in Paris.

Plate 27
KEY WEST, LANDSCAPE (1885)
Watercolor 19¼ × 13½″
Signed: LR; Winslow Homer C. S. Homer, Executor.
Philadelphia Museum of Art
Gift of Dr. and Mrs. George Woodward

This mirage-like sketch represents Key West which, when Homer visited there, was the richest city in Florida—owing largely to the wreckers who went forth from her docks in fleets of boats to rescue vessels which had been driven by waves and wind against the treacherous offshore coral reefs.

Painted during the winter of 1885-1886 while visiting with his father, Charles S. Homer, Sr., this impression offers warmth and repose. The place was one of his favorite Florida locales. Of the city itself, Homer later wrote the following to his brother, Arthur:

"I know the place quite well . . . I will once more have a good feed on goat flesh & smoke some good cigars and catch some red snappers."

For many years, this composition remained in the collection of Charles and Mattie Homer, the artist's brother and sister-in-law. It's washy and unfinished appearance have not given it a significant place or any fine accord by critics past or present. The work is excellent, however, as a typical example of Homer's initial reaction to a new setting. After he became acquainted with a place, he was able to present its finest elements.

Plate 28
HAULING ANCHOR (1885-1886)
Watercolor 21½ × 13½"
Signed: LR; Key West LL; W. H.
Cincinnati Art Museum

This watercolor presents an outstanding example of Homer's power of rendering the essentials of an impression and of giving an aspect of completeness to a vivid suggestion.

A yachtsman, he was familiar with the winds and waves. The boat here depicted is characteristic of those which plied the waters between the mainland of America and the sparsely populated little isles of the archipelago.

Broad beamed and tubby, the craft's deck load includes several pigs and a pair of horses. At the right in the distance is one of the Florida Keys' islands, with palm trees starkly relieved against the sky.

Downes reports a similar watercolor—called *Hauling in Anchor* appeared in the annual exhibition of American Art held by the Cincinnati Art Museum in 1896, and claimed this subject was found by the artist at Key West. According to Downes' further description, the paintings are probably one and the same. Hence, this work was actually completed after Homer's visit to the Island City with his father in the winter of 1885-1886 during which they also journeyed to Tampa.

Key West
Jan 1886

Plate 29
UNDER THE PALM TREE (1886)
Watercolor 11⅞ × 14⅞"
Signed: LL; Key West/Jan. 1886
LR; Winslow Homer
In a private collection.

As James T. Flexner states, Homer did not have "deep contact" with the world he painted; but here he was seeking not significance but visual pleasure. The biographer was referring directly to the tropics and perhaps the artist did not know or understand the thoughts of this young girl leaning against the palm tree. But this only made it easier for him to see her alone—purely as shape and color in a slightly secluded spot, surrounded by green foliage and native cacti, enhanced by yellow daubs to represent fruits and red streaks for flowers.

This rarely seen watercolor was originally in the collection of Thomas B. Clarke, New York, and was later sold to F. Rockefeller at an auction held on February 14-17, 1899.

Downes afforded the subject this brief but explicit description in his 1911 biography:

"*Under the Palm Tree* is an upright composition, with a mulatto girl clad in a gaily colored dress and wearing a scarf about her head and neck, leaning against the tree trunk; tropical plants fill the background."

Tampa Fla
W.H. '86

Plate 30
SPANISH MOSS, AT TAMPA (1886)
Watercolor 19¾ × 12¼″
Signed: LL; Tampa, Florida W. H. '86.
Collection of Mr. Arturo Peralto-Ramos
New York.

Jewel-like with its delicate tonal quality, this watercolor conveys the lack of pressure Homer found in Florida the winter of 1885-1886 while vacationing with his father. They did a little fishing from time to time and Homer did manage to produce a few, fine studies of the area.

The air is soft and humid, the "silvery water, reflecting the threads of eerie Spanish moss that hung like gray shrouds from the live oak," Gould describes the setting of this subject. Homer was continuing to maintain "the luminous quality of his Nassau painting; he was still experimenting with a new technique and soon would reach his stride."

He was learning to see color, "not for its own sake alone, but also in relation to the subject before him."

Rarely illustrated, this marvelously tranquil scene is one which remains true to Homer's desire to express nature exactly as she presented it to him; displaying also "his skill as a colorist in representing this earthly paradise."

Plate 31
LIVE OAKS (1886)
Watercolor 19½ × 13½"
Signed: W. H. 1886
The University of Kansas Museum of Art
Lawrence, Kansas

Described by one cataloguer as having "lethargic majesty," this expressive subject is a product of Homer's visit to the area near Tampa, Florida, with his father in 1885-1886.

There is almost an eternal quality in Homer's painting of these gigantic trees much like the giant sequoias of the west.

The artist has here realized a near perfect harmony both in coloration and in the treatment itself. Completely simplified in some respects, the tree-form masses are indicated in washes with rapid brush strokes—a few soft loose-hanging strands of gray moss standing out against a more generalized background, giving harmony and unity.

Mid-foreground in deep, rich brown splashes, the viewer can recognize the shapes of two human figures, apparently on their way to hunt carrying long-barreled guns. To the left is the form of a hound dog, its tail making a small, precise curve upward to the trees.

Plate 32
AT TAMPA (1885-1886)
Watercolor 20 × 14″
Signed: LL; HOMER '85
LR; Tampa, Fla. 1885/'86
Canajoharie Library and Art Gallery
Canajoharie, New York

In this tranquil scene of Florida's west coast, Homer's continual searching for realism is apparent—and with this, his love for truthful impressions.

In their woodland world near Tampa Bay, one bird is the roseate spoonbill, the other a common egret. Homer's innate artistic ability allowed him to closely examine these birds and project their images onto paper without making the resulting subject seem overly sentimental. Undoubtedly, while in Florida, he'd heard of the plume hunters and their massive slaughter of birds, such as those in the Everglades.

The contrast between light and dark plays a decisive role in this study. Even the blue sky reflected into the minute pool of water in the middle foreground adds interest to the interplay of colors. The feeling of movement is evident through the Spanish moss clinging to the trees, blown softly by a slight breeze.

Plate 33
REDWING BLACKBIRDS (1885-1890)
Watercolor 20½ × 13½" (sheet)
Signed in Spencerian script:
LR; Winslow Homer Florida
Philadelphia Museum of Art
Given by Dr. and Mrs. George Woodward

A careful and thorough study of Homer's tropical works offers a deeper insight into his profound interest in flora and fauna. From early childhood his mother, as an accomplished watercolorist who specialized in flowers and plants, taught her three sons to recognize and appreciate the beauty and value of nature. Her middle son clung to these lessons throughout his life.

This subject, depicting a flock of birds alighting in the tall grasses near a Spanish-moss draped tree stump and meager dead branches, is a rarely reproduced work.

Note: Other similar titles in this vein are *Song Birds, Nassau* (1885), *Two Flamingoes, Tampa* (1885) which is owned by the Brooklyn Museum of Art and *Florida Bobolinks,* probably done at Enterprise in 1890. (The whereabouts of the first and last mentioned works is not known at present.)

Plate 34
PALM TREES, BAHAMAS (circa 1888-1889)
Watercolor 14 × 16½"
The Toledo Museum of Art
Gift of Florence Scott Libbey

It seems so simple and yet this watercolor is full of meaning. The towering trees do not give the illusion of there being any wind—their shapely trunks having already been contoured by previous tropical storms—their fronds are not being blown in any specific direction. The air is dead calm—perhaps indicating the stillness before a storm. The clouds above are rain laden and billowy.

Serenity is personified in the coral-hued church spire jutting above the shorter palms. A few flecks of pale yellow in the left foreground represent the ever present flowers of foliage which Homer loved—these perhaps being croton or some other variegated shrubbery of the Bahamas.

Homer's later renderings of the palm trees carry more distinct detailing but none afford the viewer more memorable impressions of how fascinated he was by these trees. As M. G. Van Rensselaer brought out in an earlier biographical portrait of the artist, "The very essence of the tropics breathes in these new aquarelles . . . bold dashing studies of turquoise sea and blinding sun, of bright-hued plaster houses, gaudy with vines and flowers, of impenetrable luscious jungles, and wind-twisted palms."

HOMER 1890

Plate 35
ROWING HOMEWARD (1890)
Watercolor
Signed: LL; HOMER 1890
The Phillips Collection
Washington, D. C.

Painted on Lake Monroe, near Enterprise, Florida, this watercolor depicts three men in their small boat under an early evening sky. Shining through the purple mist is a red sun and its wake creates a pale orange cast onto the lake.

Downes assessed that "The water reflects the pale green tints of the upper sky, and is quiet, save for a ripple here and there." Homer's "sentiment of evening is finely expressed and broadly rendered."

Beam expressed his appreciation by writing: "Nor could photography quite convey the sense of light, and of soft, balmy air, so perfectly realized in the little *Rowing Homeward* of 1890."

Note: A companion subject, *Three Men in a Boat* (1890) is a most unusual work; "its draftmanship," according to one cataloguer, "is superficial . . . the whole weight of the work rests on a wash technique . . . the men and their rowboat are almost as fluid as the reflections . . . Experimentally turned on its side, a photograph has been mistaken for a contemporary non-representational painting." This watercolor is in the collection of Colby College Art Museum, Waterville, Maine.

HOMER 1890

Plate 36
ST. JOHN'S RIVER (1890)
Watercolor 20 × 14″
Signed: LR; HOMER 1890
The Hyde Collection
Glens Falls, New York

This beautiful composition was painted by Homer during his winter stay at Enterprise, Florida and has been appraised by one Homer student as the most successful of the period. The beauty of the lagoon is disturbed only by the two men in the boat, one whose fishing line is pulled taut by a catch. In the background the trees are densely hung with Spanish moss—the foreground water is filled with shimmering, rippled reflections. Homer has treated the background very simply in terms of two predominant masses—these being two trees so heavily hung with moss that they stand out just by virtue of their greater detailing. The overall view is sufficiently simple and the viewer is attracted singularly to the handsome patterns made by the hanging gray moss.

Homer rendered the water with swift brush strokes—lightening the reflections and going darker into the background. A faint, tiny sweep of his finest brush suggests the tight fishing line. The expressive color scheme is warmly tropical. Even the grays are alive.

Previously titled *Fishing in the Lagoon,* this work was among the collection of paintings which came into the possession of Mr. and Mrs. Charles S. Homer following Homer's death in 1910.

HOMER 1890

Plate 37
WHITE ROWBOAT,
ST. JOHN'S RIVER (1890)
Watercolor 20 × 14″
Signed: LR; HOMER 1890
Cummer Gallery of Art, Jacksonville, Fl.

Among Homer's favorite Florida fishing spots was along Lake Monroe on the St. John's River, 122-miles up river from Jacksonville. He came here by steamboat to stay at Enterprise, a small community across from Sanford. The locale boasted fine luxury hotels and posh continental-inspired eating places but Homer preferred the lesser known and more economical hostelries, such as the Live Oak House or Lemon House which offered accommodations for $2 per day.

The original watercolor of this subject is perhaps the most vivid of the paintings Homer did at Enterprise in the winter of 1889-1890. The treatment of the water is simple yet vigorous—a few strokes suggesting the whole with its reflections and shimmering quality. Seated in the white boat are two figures, one tugs on a line, silhouetted against a mass of brilliant tropical foliage. The rapidity of brush strokes forecasts Homer's remarkable later works.

Plate 38
PALM TREES,
ST. JOHN'S RIVER, FLORIDA (1890)
Watercolor 19¾ × 13¾" (within mat)
Signed: LR; Winslow Homer 1890
Courtesy The Kennedy Galleries

This painting from Homer's trip to Florida during the winter of 1889-1890 bears a striking resemblance to an 1890 roundel photograph, now in the Homer Collection at Bowdoin College. "Winslow Homer and Guides, Homosassa River, Florida," another photograph, is so similar it may be doubted whether the locale designated on the second photo is correct, since Homer did not go to Homosassa until 1904. (Both photographs are shown in the preceding text.)

Philip C. Beam feels that this watercolor "reminds us of the intoxicating beauty of moonlight over tropical water." This work and *The White Rowboat, St. John's River* were long owned by Charles L. Homer, the artist's nephew and son of his brother, Arthur.

Palm Trees, St. John's River, Florida offers "striking evidence of the superiority of artistic seeing and accenting over the more impersonal eye of the camera," Beam further attests.

Plate 39
THE PALM TREE (1890)
Watercolor 15 × 21″
Signature: LR; partially cut off.
Fogg Art Museum, Harvard University
Bequest of Grenville L. Winthrop

Homer's tropical paintings were basically horizontal—capitalizing on their expansive width. Here, however, he offers pictorial tension with vertical elements in the arching palm fronds, while a small amount of horizontal stress is found in the fallen, dead fronds and dried coconuts in the foreground. He has kept the background more generalized and completely simplified with rapid brush strokes.

Mass forms are indicated but are washed in and flowers become mere dabs of color. There is a unity and harmony evident which Homer had not attained before.

Homer's fascination for the palm tree led him to paint nearly a dozen watercolors in which this specie of tree—singly, paired or in a cluster—is the main theme. A horticulturist by instinct, he expressed his profound admiration for all living things with every brush stroke and careful selection of color. The white clouds behind the tree give a diffused light. His artistic veracity of photographic beauty offers the dominant quality of the painting.

Plate 40
PALM TREE, NASSAU (1898)
Watercolor 15 × 23½"
Signed: LR; HOMER
The Metropolitan Museum of Art
Amelia B. Lazarus Fund.

Homer kept this watercolor, from his superb series of southern subjects, because he considered it to be among his best works.

The painting and eleven others were acquired by the Metropolitan Museum of Art following Homer's death. Downes asserted that the "museum made an excellent selection, and Homer on his part manifested his customary sagacity in setting aside these works for the permanent collection of the leading art institution in America."

A catalog description for a February 1911 memorial exhibition reads as follows: "*Palm Tree, Nassau* represents a lofty coco palm bent by the wind, and several smaller palms beyond. In the background is a deep blue sea, a narrow strip of land and a white lighthouse."

This extremely popular watercolor has probably been reproduced more often than any other of Homer's tropicals, aside from the oil painting of *The Gulf Stream* also in the Metropolitan's permanent collection. It is just one of his numerous renderings featuring Hog Island Lighthouse, the landmark which to this day greets visitors entering the harbor of New Providence Island.

Winslow Homer 1890

Plate 41
UNDER THE COCO PALM (1898)
Watercolor 20½ × 14⅜"
Signed: LR; Winslow Homer 1898
Fogg Art Museum, Harvard University
Louise E. Bettens Fund

Homer might have seen this captivating subject from a window of the house he rented in Nassau while vacationing there in the winter of 1898-1899. The young black boy, comfortably seated among fallen palm fronds, is contentedly drinking coconut milk.

Also titled *Negro Under Palm Tree,* this charming watercolor was given by Homer to Henry Wingate Stevens in payment for his supervision services during the building of a cottage near the eastern tip of Prout's Neck in 1901. Sometime later, Stevens sold this painting to M. Knoedler, the art dealer with whom Homer had close contacts, and the proceeds from the sale were used to put his son, Theodore M. Stevens, through medical school.

Plate 42
NEGRO CABINS AND PALMS (1898)
Watercolor 21 × 14⅜″
Signed: LR; Winslow Homer 1898
The Brooklyn Museum

Viewed offshore, perhaps from a turtle boat, this is probably a scene Homer saw on his second excursion to the Bahamas. He again found the mood of the natives contagious. They were "friendly in an offhand fashion, carefree, given to easy laughter, indolent and industrious by turn, depending on the time of day and the circumstances," Jean Gould records.

Homer was, of course, elated that nothing seemed to have changed: "the joyous primitive life of the natives, the long, narrow stretches of beaches, and the brilliant tropical landscape unhampered by hotels" was still as he recalled.

In his early sixties, the artist gloried in painting these pictures—reveling in sun and sand and the tropical sea. Each day found him out early to capture the rapidly emerging pageantry before him. He was working in a kind of ecstacy with a swift, sure touch. He depicted the palms loosely and in dark, gray-green shades against the pale luminous, possibly early-morning, sky. The tree forms, in irregular patterns, tower over the small cabins and heeled over boats. The craft had probably been upturned on the beach for badly needed repairs. At the left foreground a cluster of three men give careful examination to the stern and keel of one boat. A tiny fleck of red on the right foreground represents another native figure beside another boat. The pink-tinged beach is lapped gently by a soft blue sea.

Previously titled *Key West, Negro Cabins and Palms,* the location designation has been dropped since modern critics and biographers are in agreement that the artist did not visit that Florida island city at the time this subject was reportedly conceived and completed.

Winslow Homer
with compts to
Mrs Delos McCurdy 1899

Plate 43
NASSAU: BEACH WITH CACTUS (1899)
Watercolor 21¼ × 11″
Signed: LL; Winslow Homer with compts to/Mrs. Delos McCurdy 1899
Indianapolis Museum of Art
Mary B. Milliken Fund

An air of mystery surrounds this interesting watercolor; the title is somewhat misleading since the foreground is not a beach at all but the top of a plastered wall into which broken dark-brown jagged glass bottles have been inserted and the cactus are probably Century Plants. (Upon appearing in a catalog produced by the International Arts Guild of the Bahamas for a 1962 exhibition at Government House, Nassau this subject was afforded the title *Nassau: Beach with Century Plants.)*

The inscription has also created a certain amount of speculation; as to just who Mrs. Delos McCurdy was, and what prompted Homer to present her with this fine almost oriental theme. We do know, however, that the artist was in the habit of giving his paintings to persons who had done special services and favors for him. Perhaps Mrs. McCurdy owned the cottage where Winslow stayed while in Nassau on his second sojourn the winter season of 1898-1899.

To our knowledge, this is the first time a color reproduction has ever appeared of this painting.

Plate 44
HURRICANE, BAHAMAS (1898)
Watercolor 21 × 14½"
The Metropolitan Museum of Art
Amelia B. Lazarus Fund.

Formerly titled *Tornado, Bahamas,* this watercolor is one of twelve acquired by the Metropolitan Museum of Art from Homer's estate following his death September 29, 1910.

The work depicts coconut palms swaying in a gale above a small group of roof tops. A glimpse of dull green sea is to the left while heavy storm clouds fill the sky above.

Earlier, while on his 1885 visit to the Bahamas and Cuba, Homer painted *Windstorm, Bahamas* (also known as *Wind Storm in the Tropics).* This work is privately owned and was illustrated in the *American Magazine of Art,* October 1936, accompanying an article by Forbes Watson, commemorating the 100th anniversary of Homer's birth. Of the watercolor *Windstorm, Bahamas* and *Palm Trees, Nassau,*—not to be confused with *Palm Tree,* dated 1898 Nassau (plate 40)—both 1885 works, Philip C. Beam says this: "... he had shown the trees bent like taut bows under the violent pressure of the gale." Describing *Hurricane, Bahamas* he states: "Now, in 1898, he combined his thoroughly mastered elements into a piece charged with mood but rigorously ordered . . . The storm, more accurately called a hurricane, is indicated with great power by the striking value contrasts and the somber low intensity tones; the palms, expressive as they are, are generalized to suggest the nature of all palms without representing any particular ones. It is a wonderfully satisfying and moving example of a maximum effect achieved with a minimum of means . . ."

HOMER

Plate 45
RUM CAY (1898-1899)
Watercolor 21⅜ × 14¹⁵⁄₁₆"
Signed: LR; HOMER
Worcester Art Museum, Worcester, Massachusetts

Monumental in its simplicity, this painting relates the age-old story of the fight for survival—a story Homer chose to retell many ways over the years.

The huge sea creature probably had gone ashore to lay her eggs and was attempting to escape the agile native. Capture is, however, inevitable. Homer witnessed such events as he was often invited to go along on the turtle boats harbored at New Providence Island. On one such trip he went to the Harbour Islands near Eleuthera, northeast of Nassau.

Jean Gould in her biography of the artist terms the black's "realistic movement" to be "as rhythmic as a dancer." The work was given the title *Rum Cay, Bermuda* in William H. Downes' 1911 biographical study, but Homer denoted the watercolor as *Turtle Cay,* beside a tiny sketch in his Day Book of 1902. When, by whom, or why the title was altered is unknown.

Beam adds that " 'cay' like the Anglicized 'key' is from the old native Taino word for island, and there is a Turtle Island off the coast of Georgia. This, or a similar location in the Bahamas, probably gave Homer his original title. He may have rechristened it later; or his brother Charles may have named it after Winslow's death, from Rum Cay, a small island east of Nassau."

HOMER 1898

Plate 46
THE TURTLE POUND (1898)
Watercolor 21 3/8 × 14″
Signed: LR; HOMER 1898
The Brooklyn Museum
A. T. White and A. A. Healy Fund

This watercolor for many years was erroneously called *Turtle Pond.* A small thumbnail sketch of the subject appears in Homer's Day Book for Saturday, July 5, 1902. His handwriting adds the notation: *"Shall send to Knoedler"* . . . Then, a listing of ten watercolors follows with *Turtle Pound* denoted as number five.

This subject became the most popular of his 1898 series. The painting, in no way, gives one the impression or feeling that the models were posed, but as Beam relates, "that Homer caught them unselfconsciously at their accustomed tasks." Just the upper torso of a handsomely-proportioned Bahamian black and the head and hands of another are skillfully utilized by Homer. A bronze-toned shoulder and darker shade head are starkly silhouetted against the bleached, natural wood of the pen or krawl in which the turtles are being placed after their capture. (Thus kept alive and properly fed, the reptiles could provide the natives with fresh meat for months on end.)

Dec 31 1898
Nassau

Plate 47
A WALL, NASSAU (1898)
Watercolor 21½ × 14¾"
Signed: LR; Homer Dec. 31, 1898 Nassau
Metropolitan Museum of Art
Amelia B. Lazarus Fund

The subject depicts a white plastered wall atop which broken glass bottles have been cemented in an effort to keep intruders out. The vivid red poinsettias against the pale sky are a colorful challenge.

Donelson Hoopes found this watercolor from Homer's second trip to the Bahamas to be "curiously close to the technique he first employed there. A certain dryness which is characteristic of his earlier Bahamian watercolors prevails." The subject matter of a simple stone wall, with its "singular lack of decoration," serves elegantly as a foil for what Hoopes believes to be "the highly colorful frangipani." Plumeria, the Latin name for frangipani, blooms in the warm spring and early months of summer; its blooms are generally white or bold pink. Poinsettias would definitely have been in full blossom in December; since this plant is customarily considered a Christmas flower.

Homer deliberately selected this subject, perhaps "setting up a challenge for his abilities. To paint a nondescript wall as an important element in a composition demands an enormous assurance," Hoopes asserts. He goes on to point out that John Singer Sargent, a rival artist of Homer, was painting similar walls about the same time on the island of Corfu in the Mediterranean.

Revealing itself "slowly, despite its apparent simplicity," *A Wall, Nassau,* presents us with a "study in contrasts, balances, and tensions: the contrast of neutral and bright color, the balance of the doorway's exaggerated off-center position against the dominant emptiness of the wall; and the tension between the doorway and the distant sailboat."

The poinsettia blooms, exploding "against the ultramarine sky is a transcendent visual experience," wrote Hoopes in his final appraisal.

HOMER

Plate 48
THE WATER FAN (1898-1899)
Watercolor
Signed: LR; HOMER
The Art Institute of Chicago
Gift of Mrs. John A. Holabird

The Water Fan, a creation of Homer's second visit to the Bahamas, possesses exquisite clarity and strength. Just as he had hoped, the blacks were still as receptive to his painting them as they had been on his previous trip.

This painting depicts a fisherman using a glass bottomed bucket to see treasures in the shallow waters near shore. One such "find" is the delicate peach-toned sea fan already laid carefully into his boat. The man's face and hands are handsomely detailed.

Goodrich remarks that "The superb beauty of the black bodies and the spectacle of this free life in such a setting gave these works, with all their realism, that pagan spirit we call Greek."

Note: A companion painting, *The Sponge Diver, Bahamas,* owned by the Museum of Fine Arts, Boston, is also a strong and colorful but simple watercolor. It shows the powerful back of a Negro diver vividly silhouetted before the gleaming stern of the skiff and the emerald sea.

HOMER

Plate 49
NASSAU (Formerly titled BERMUDA) (1899)
Watercolor 21⅜ × 15″
Signed: LR; HOMER LL; Jan. 1, 1899
The Metropolitan Museum of Art
Amelia B. Lazarus Fund

This and many of his other tropical subjects were retained by the artist for many years, some hanging on the walls of the small cottage he had built and rented out to tourists during the summer, when Prout's Neck became increasingly popular as a resort area.

This admirably crisp watercolor was described in the following manner in a catalog for the Metropolitan's exhibition in February 1911: "On the white beach in the immediate foreground are three rusty cannons; deep blue sea beyond. A sail boat, manned by two Negroes, is near the store, and several other vessels are farther out. In the distance is a line of brown shore."

Homer found this scene while making his second sojourn to the Bahamas in the winter season of 1898-1899 and, as previously, this sunny region was for him both an actual world and a realm of fantasy.

Here, in full force, he achieved "his highest, brightest coloring," Flexner asserts, adding that Homer "did not, however, depend for brilliance on the elimination of dark hues, but traveled the whole gamut from black to the greatest brilliance, often in a single bold painting." *Nassau* is, indeed, a fine example of this critic's evaluation.

Plate 50
SLOOP, NASSAU (1899)
Watercolor 21½ × 15″
The Metropolitan Museum of Art
Amelia B. Lazarus Fund

Beam assesses this composition as having "unification in its rhythmically patterned curves and a color scheme amazingly rich in both an abstract and a descriptive way."

Here, again, Homer returns to his old love . . . a graceful boat at anchor. In this, as in his other similar tropical watercolors, Homer was "continually simplifying, reducing" his themes . . . "more and more to their essential lines and masses."

Donelson F. Hoopes, one of Homer's more recent biographers, places this watercolor in with the artist's Bermuda studies; stating these were "freer and more lyrical in color than those of even a year before in the Bahamas." He feels there is an "ease of drawing and a certainty in the placement of each color passage that speaks eloquently of Homer's sense of personal well being."

The Metropolitan Museum of Art's catalog in 1911 for the Homer Memorial Exhibition described this work as: "—a white sloop seen from the stern, where a reddish rowboat is tied. The water is green and blue. Aboard the sloop are two negroes. The sails hang in windblown swirls. Clothes are hung out to dry on the boom. There is a small boat on the right, and a strip of brown shore at the horizon."

Most recently John Wilmerding, Dartmouth College art professor, put this watercolor in its proper location classification, changing the designation from Bermuda to the Bahamas. And, in comparing this subject with *Nassau, 1899* one realizes the close similarity not only in the sailing craft illustrated in each, but also in the brown-toned land form on the horizon.

Note: One of his earlier works, *Market Scene, Nassau* (1885) depicts two sailing craft, two rowboats, and six natives exchanging a pair of trussed chickens from one boat to the other.

Plate 51
LIGHTHOUSE, NASSAU (1899)
Watercolor 21⅛ × 14½″
Signed: LL; Nassau - '99 LR; Winslow Homer
Worcester Art Museum
Worcester, Massachusetts

Homer here depicted a native conch diver of the Bahamas waving to friends in another boat, represented by a mere splotch of black on the horizon near the storm warning flag and lighthouse on Hog Island. The seas are choppy as one denotes from the wave in the foreground which seems about to swamp the small craft.

Note: *The Sponge Diver, Bahamas* (1889) is a similar effort using a simplified design built around a single figure.

Plate 52
WEST INDIA DIVERS (1899)
Watercolor $21\frac{1}{4} \times 14\frac{1}{2}$"
Signed: LL; HOMER, Nassau '99
The University of Kansas Museum of Art
Lawrence, Kansas

The majority of Homer's Bahamian works depicted the ordinarily peaceful day-to-day life of the tropics with an easy intimacy, capturing the relaxed quality of work in those beautiful surroundings. This watercolor subject is just such a scene.

Being the opposite of spectacular, its charm lies in the complete absence of pose or affectation on the part of the models portrayed. Handsomely painted are the perfectly modeled head of the man, the warm browns of the young boy's torso, the rich tropical background and the limpid quality of the shallower waters along the beach. A lifetime of experience lay behind Homer's deceptively simple rendering which makes human behavior seem so natural and watercolor so easy.

Homer loved being amongst the men of the West Indies who relied upon the seas for their living. The sea and its inhabitants were an integral part of their environment. Homer studied their ways very carefully and he was always a welcome guest with any fishing crew. Everywhere he traveled, he soon acquired favorite guides who saw to it that he found exactly the right setting he desired for his work.

HOMER

Plate 53
AFTER THE HURRICANE (1899)
Watercolor 21⅜ × 14⅜″
Signed: LL; HOMER '99
The Art Institute of Chicago
The Ryerson Collection

The artist's own Day Book of 1902 gave the title *After the Tornado, Texas* to this distinctive painting and Philip C. Beam in his book, *Winslow Homer at Prout's Neck,* surmises that the subject may have been suggested to the artist during one of his visits with his brother, Arthur, at his winter home in Galveston, Texas. Lefacaido Hearn's "Chita: A Memory of Last Island," the tale of a tremendous hurricane on the Gulf coast, first appeared in *Harper's New Monthly Magazine* in April, 1888. Later this writing was published in book form by Harpers.

Jean Gould commented: "After a violent tropical storm the sight of one of these strong boys lying dead on the beach, his splintered sloop nearby at the spot where the storm had tossed them, evoked the telling, dramatic *After the Tornado.*" Lloyd Goodrich assesses this work as representing "Homer at the height of his creative powers" and that here "he has translated objective vision into pictorial terms" and has achieved a composition which is "powerful in expression and design."

HOMER

Plate 54
SHORE AT BERMUDA (1899)
Watercolor 21 × 13⅞"
Signed: LR; HOMER
The Brooklyn Museum

This tranquil scene offering a wide expanse of bright-blue sea is perhaps one which Homer was afforded from the window of his rented bungalow. Previously this subject was incorrectly titled *Shore at Nassau* and frequent reference to it with that title appears in many biographical works.

Another of Homer's similar watercolors completed in Bermuda is *Rocky Shore,* which carries an inscription in his own handwriting on the reverse side of the painting that reads: "Homer—Landing Bermuda, 1900". This subject has been described as depicting a bright blue sea, paling to blue-green between gray cliffs; two red coated soldiers are on white beach to the right.

Similar in concept is *Natural Bridge, Nassau* (1898) which again brings the problem of proper location designation since it is a well-known fact that many of these tropical works were retained by the artist for his own viewing and enjoyment and that he guarded them almost jealously from public view as the critics had not always been thoughtful in handling their commentaries. After Homer's death, his brother afforded them further identification at his own discretion.

Plate 55
FLOWER GARDEN AND BUNGALOW, BERMUDA (1899)
Watercolor and pencil 21 × 14″
Signed: LR; Winslow Homer/Bermuda 1899
The Metropolitan Museum of Art
Amelia B. Lazarus Fund

This vibrant subject was one of fourteen of Homer's choicest "Scenes from the Bahamas and the Bermudas," which he himself selected and offered to Knoedler's for exhibition and sale at $200 each in 1902. The year before, twenty-one tropicals had been exhibited at the Pan American Exposition in Buffalo and Homer received a gold medal.

Jean Gould states that the artist went to Bermuda in December, 1899, "where he rented a small bungalow beside the deep-blue sea, set off by the riotous color of a tropical flower garden—scarlet hibiscus, red and yellow cannas, small banana palms, with their plump decorative fruit, bordered by plants of variegated foliage—and one of the first watercolors he painted pictured the place where he stayed."

At a corner of the house, mid-right section of the painting, the minute figure of a small black girl is just discernible and appears to be running toward the flowers in the foreground. This is one of the rare instances where a human figure appears in Homer's Bermuda subjects.

The handsome architecture of the house with its yellow walls and white cement roof stands out against the blue bay and the shoreline to the right is dotted with white buildings. Beam feels that this watercolor "stresses the forms and colors of the vegetation." It is a painting often reproduced and favored by art lovers throughout America who recognize it as being a scene quite typical of Bermuda.

Plate 56
HOUSE AND TREES (1899)
Watercolor 21 × 13⅜"
Signed: LR; Sketch - W. H.
The Brooklyn Museum

Note: A similar subject, *Palm Trees and Houses, Nassau,* which bears the initials W. H. in lower right section, is in the collection of the Museum of Fine Arts, Boston, a bequest of John T. Spaulding. Its authenticity is reported "doubtful."

Previously titled, *House and Trees in Nassau,* this subject was long associated with Homer's other Bahamian works. Careful examination reveals, however, that "the houses represented are characteristic of those found on the island of Bermuda," according to Donelson F. Hoopes. He also considers it "an unfinished work, for the foreground, with its incomplete figures, appears to be merely a first wash which was never built upon."

The painting "captures the charm of the relaxed and balmy days" in a peaceful island haven—whether in the Bahamas or Bermuda. Hoopes feels "it is tempting to speculate about what the figures are intended to be doing" crawling across what appears to be newly turned earth. Homer has "worked toward unification of color by introducing the same red into the trees" as is in the foreground earth. This is a practice he also used occasionally in his seascapes.

Bermuda
W.H. 1900

Plate 57
BERMUDA (1900)
Watercolor 20½ × 13½″
Signed: LR; Bermuda W. H. 1900
Philadelphia Museum of Art
Given by Dr. and Mrs. George Woodward

Homer provides us a gracious access to a secluded stretch of Bermuda shoreline. The feeling of serenity is very strong. Homer arrived in Hamilton, Bermuda in December, 1899, rented a small bungalow and began a series of land and seascapes; some showing the riotous colors of tropical flower gardens, the neat pastel-hued houses and the trees and other verdant foliage. His technique was free and full.

Human figures rarely appear in Homer's Bermuda studies. His subject matter more often concerned various offshore islands, spits of sand jutting into azure water, or curves of beach lined with palms and houses in the distance. Homer so prized his Bermuda series that he was reluctant to place them on exhibition.

W.H._1900

Plate 58
NORTH ROAD, BERMUDA (1900)
Watercolor 20¾ × 13½″
Signed: LL; W. H. 1900
Collection of Mrs. John Wintersteen

This subject from one of Homer's trips to Bermuda appeared incorrectly titled in a catalogue of the 1960's—an incident which has occurred frequently since many times the artist gave similar titles to different paintings or even conferred the same title to watercolors and subsequent oils.

The placid seascape with the Atlantic appearing like a mill pond and a steamship advancing across the horizon towards its next port of call is one of the sights typical to Bermuda. Homer's technique was freer here and he spent his attention singularly on the oddly bent coastal trees. The artist seemed at ease with few worries and enjoyed his painting vacation.

Note: Another more vigorous subject is *The Coming Storm* painted in 1901. His study *Gallow's Island, Bermuda* (circa: 1899-1901) is similar in its ability to express the fine luminosity of tropic sea and sky. This handsome signed study sold for $65,000 in November, 1972.

Plate 59
CORAL FORMATION (1901)
Watercolor 30⅜ × 14″ (sight)
Signed: LR; Winslow Homer 1901
Worcester Art Museum
Worcester, Massachusetts

This watercolor, according to Philip C. Beam, carries a title which was added by Homer's brother Charles, following the artist's death in 1910. Beam asserts it could surely be a misnomer and further explains that "the formations are not coral but stratified and water-eroded limestone (Karst is the geological term)" and are common to Cuba and the West Indies, but extremely rare in Bermuda . . . "hence my guess that the Bermuda painting is actually probably the Bahamas," Beam continues.

Note: *Natural Bridge, Nassau* (1898) and *Rocky Shore, Bermuda* (1900) give further strong evidence of Homer's fascination with rocks and strata during this period.

HOMER
1901

Plate 60
BERMUDA SETTLERS (1901)
Watercolor 20⅞ × 14″
Signed: LR; HOMER/1901
Worcester Art Museum
Worcester, Massachusetts

This unique painting was listed in Homer's Day Book as No. 9 of a group of watercolors he sent to Knoedler's in 1902; painted the year previous, the artist at that time called it *Bermuda Pigs.* Later, in 1923 it was reproduced as *Dans le Jungle, Florida* in the *Revue de l'Art.*

Beam reports that Homer did a preliminary sketch for this subject, "showing several razorbacks, one of whom faced the viewer with his prominent red snout." The artist, calling the visitor's attention to the sketch, would later pull off the hog's nose (actually a red tag from a plug of Lucky Strike chewing tobacco) and move it over to the horizon. When the visitor looked at the picture again, the tag had become a scarlet setting sun.

This episode was just one example of Homer's keen sense of humor. Downes earlier stated that the artist's interest in the native pigs came from his introduction to some of these animals being carried on deck of one of the Bahamian boats trading between the mainland and the sparse settlements of the archipelago; probably some of these were descended from a breed brought from Africa to the Bahamas many years before.

Key West

Plate 61
KEY WEST (circa 1903-1904)
Watercolor 20¾ × 13¼″
Signed: LR; Key West LL; W. H.
Fogg Art Museum, Harvard University
Gift of Edward W. Forbes.

In his tropical watercolors Homer quickly painted what was directly before his eyes. Frequently, as in the case of his Key West marine series, he might coordinate his subjects. He usually separated the conception, inspiration and finished product by only a few hours instead of the weeks and sometimes months as was the case with his oil paintings.

Prior to his trip to the Old Island city, some one hundred and sixty miles from the mainland of Florida, Homer had written that he had an idea for some work. He had a lifelong interest in boats, and the working craft of any harbor—whether in Nassau, Gloucester or Prout's Neck—offered him inspiration.

The nearly matching tones of blue in this painting are typical of the appearance of water and sky on a calm, smooth day in the shallow areas off Key West. Sails furled, these vessels offer a scene of repose—such as a Sunday morning might afford. How different an atmosphere here, compared with the bustling harbors of Tynemouth and Gloucester with all their activity and human evidence.

The artist was 68-years old and he relished the quiet and seclusion afforded him in Key West. More and more his works were devoid of people. Nature was of its usual utmost importance to him.

Dec 22
HOMER

Plate 62
STOWING SAIL (1903)
Watercolor 21$^{13}/_{16}$ × 13$^{15}/_{16}$″
Signed: LL; Sketch from Nature
Dec. 22, 1903 HOMER
The Art Institute of Chicago
Ryerson Collection

This particular watercolor usually bears a location tagline, *Bahamas.* This is purposely deleted here since the feeling of many of Homer's biographers and critics is that this was erroneously added by the artist's brother, Charles S. Homer, who was executor for the artist's estate following his death.

The subject matter is more compatible with the other marine studies Homer did in Key West in December 1903 and January 1904. The foreground area depicts the sun brilliantly reflected on the soft azure waters, while on the far right horizon line, the slight form of an island takes shape. (This landform appears similarily in *The Conch Divers* and *Hauling Anchor.)*

The painting's overall appearance is transparent and the feeling is profoundly cool and airy. We also see here vivid evidence of Homer's use of red, since he deftly managed to inject three splashes of it in strategic positions.

Note: *Weighing Anchor,* another of Homer's Key West subjects of 1903, portrays a two-masted craft with her sails flapping gently in the trade winds while the anchor is being hoisted. Privately owned, this watercolor offers livelier tones of blue—it also is signed and dated in the lower left corner.

LIZZIE

Plate 63
FISHING BOATS, KEY WEST (1903)
Pencil and watercolor 21½ × 13½″
Signed: LL; HOMER 1903 LR; W. H.
The Metropolitan Museum of Art
Amelia B. Lazarus Fund.

Representing another dazzling flood of southern sunlight this handsome study of boats in the harbor at the southernmost city of the continental United States, conveys the impression of airiness and brilliance so characteristic of the tropics.

Philip C. Beam offers the theory that the subject is basically a portrait of a single boat, in "the simplest of designs; a mere hint of modeling" to offer "the feeling of form, solidity, and distance." As in *Stowing Sail,* the watercolor "is a study in washes of unsurpassed freshness." And, here, too, we find small touches of bright red which "enliven the otherwise cool palette."

Biographer Beam feels that "even in these simple, apparently ingenuous watercolors, Homer's incredibly effective use of color is undeniable evidence that he is one of the most knowledgeable colorists in all modern art."

Throughout his life, Homer "regarded watercolor and drawing as ideal media for unpretentious subjects . . ." Oil, to him, was "the logical medium for bigger ideas. Yet, curiously, it was in his drawings and watercolors that he achieved some of his fullest realization of three-dimensional form . . ."

Plate 64
FISHING BOATS, KEY WEST (1904)
Watercolor 21½ × 13½″
Signed: LL; Key West 1904/
LR (on boat stern); HOMER
Worcester Art Museum
Worcester, MA.

The strong, essentially blue hues of this rendering give it a deeply moody feeling. Homer partially relieved the painting's somber coloring with slight splashes of gold and his familiar touches of bright red.

He loved the clean lines of the boats in Key West's bustling little harbor; numerous masts cleave the dark sky and the waters below offer slight movement in perpetually undulating swells. Turbulence threatens in heavy clouds and high, upper-atmosphere winds.

Not one of the more popular subjects of his marine series, this painting should not be overlooked, however, for it demonstrates the variety of weather which Homer loved to paint.

Plate 65
HOMOSASSA JUNGLE IN FLORIDA (1904)
Watercolor 21½ × 13¾″ (sight)
Signed: LL; HOMER 1904
Fogg Art Museum, Harvard University
Gift of Mrs. Charles S. Homer

At the small village of Homosassa on Florida's Gulf coast, Homer found a fisherman's paradise. He relished the quiet isolation and excellent sport this secluded spot afforded him.

This deep-hued watercolor depicts the thickly-forested shores along the Homosassa River, which is fed by the gigantic Homosassa Springs further inland. Homer's deep admiration for the still untamed, natural beauty he found here is richly expressed. Note how the calm waters reflect the swamp grasses and overhanging branches above the shoreline.

Note: Other subjects from this period are: *Red Shirt, Homosassa, Florida; Homosassa River,* and *The Turkey Buzzard*

1904
HOMER

Plate 66
THE SHELL HEAP (1904)
Watercolor $19^{11}/_{16} \times 14''$
Signed: LR; HOMER: LL; 1904
The Brooklyn Museum

The painting's setting is along the famous Homosassa River. Winslow Homer wrote to his brother, Arthur, in January 1904:

"Delightful climate here about as cool as our September—"

The community where he stayed during his sojourn was sparsely settled, with perhaps less than 150 inhabitants calling it their permanent home. These were mostly natives, fishing guides, a few retirees from the North and a family or two that had come south from Cedar Key when the pencil factories there were forced to close after the cedar forests were depleted. Homer arrived here by boat from Key West on the regularly scheduled fleet of boats that plied Florida's western Gulf coast.

The painting's lower left corner displays a small pile of sea shells. The settlers, as the Indians of the area had done previously, would select a certain spot and there dump their oyster, scallop and other shells. These mounds or middens grew over the years into enormous piles. Homosassa River and Springs continue to be favorite areas for fishing enthusiasts. Many miles of the Homosassa's coves and inlets still resemble the wilderness which Homer captured 68-years ago.

Plate 67
THE TURKEY BUZZARD (1904)
Watercolor $19\frac{3}{4} \times 13\frac{7}{8}$"
Signed: LL; HOMER 1904
Worcester Art Museum
Worcester, Massachusetts

This famous watercolor is, according to Beam, another product of Homer's sojourn at Homosassa, Florida. The "principal feature is a grove of trees which grew beside the river bank and stood out strongly against the sky. That is the way visitors to the tropics remember the tall palms, and Homer, as always, captured the characteristic. Trees always interested him wherever he traveled. In the north woods it was the evergreens observed and depicted from every conceivable angle, but always related to their environment; in the south it was the tufted palm bent under a hurricane or swaying with the trade winds. In keeping with its humid serenity, the sky in *The Turkey Buzzard* is slightly overcast and the trees are deeper-toned than the flashing colors in the sun drenched *Shell Heap*. The perfect touch is the turkey buzzard which floats lazily in the still air on its scavenger's vigil. It was a type of touch which the mind of Homer employed often to convey his conception of the fundamental facts of the natural world. Again he included himself and his guide fishing from their boat, but small and distant so as not to distract from the grove of palms or the buzzard. Each goes about his business with uncalculated aplomb, making Homer's statement utterly true to life."

Plate 68
IN THE JUNGLE, FLORIDA (1904)
Watercolor 19⅝ × 13⅞″
The Brooklyn Museum

For this watercolor, according to Philip C. Beam, Homer attempted "to give the subject more meaning by including a forest dweller, but the final effect is confusing." Beam continues, "Gradually, laboriously, he achieved the simplification of this complex and psychologically overpowering subject (the jungle) to a degree"—retaining the feeling of the original earlier watercolor, *Florida Jungle,* painted in 1885, but allowing him "to arrange its elements into an acceptable design." This subject belongs to his Homosassa River series but is a distinct variation from the others.

DIAMOND
HOMER

Plate 69
DIAMOND SHOAL (1905)
Watercolor 21½ × 13½″
Signed: LL; Diamond Shoal LR; HOMER 1905
IBM Corporation

Philip C. Beam wrote that Homer "had done little work during 1905, and under the pressure of inspiration painted *Diamond Shoal.*" This is considered one of his most impressive final works. It demonstrates wonderful originality in its emphasis of the "headlong, lifting surge of the schooner by opposing it at right angles to the distant, bobbing lightship." One definitely feels the dramatic tension of this episode which Homer witnessed many times on his voyages by steamship from Norfolk to Florida and the Bahamas, passing Diamond Head Lightship 17-miles offshore from Cape Hatteras, "the Graveyard of the Atlantic." In an exhibition catalogue, Beam states the following: ". . . he perceived its dramatic and pictorial possibilities and recorded them with a seaman's eye, giving us every requisite for proper interpretation. The bobbing lightship and urgent activity of the sailors tell us that a gale has suddenly aroused the seas. In quick response the seamen have dropped the flying jib to the bowsprit, hoisted the mizzen sail to point the bow into the wind, and are frantically trimming the staysail. The mainsail has been hurriedly lowered but left unfurled. Quietly dramatic, these signs of expert seamanship are a part of that timeless mariner's knowledge which Homer encompassed with authority. Yet, thanks to another side of his genius, a layman can enjoy the consummate use of foreshortening to heighten the thrust of the schooner through the waves, for no artist in modern history ever made boats ride on water with a surer instinct than his. Here, then, is not a spectacular ocean tragedy, but a long-range view of the perils of the sea. A man of comprehensive vision, Homer was mindful of both sides."

Plate 70
THE WRECKED SCHOONER (1908)
Watercolor 21 × 14½″
City Art Museum of St. Louis

The highly destructive power of the sea and its ever-present threat to human life, was an element which Homer witnessed and returned to as a painting subject many times throughout the years. *Gulf Stream* and *After the Hurricane* were but two tropical variations of this continuing theme.

Beam assesses, "It was therefore nearly inevitable for Homer to return to this field once more." The spectacular destruction of a super-schooner named the *Washington B. Thomas* furnished Homer the subject, when he witnessed the ship's tragic aftermath during the summer of 1903. Constructed at the Watts Shipyard in Thomaston, Maine. The *Washington B. Thomas* was returning from her maiden voyage to Norfolk, Virginia, when she became becalmed at night in a dense fog near the mouth of Portland Harbor. Since no harbor pilot was available at that hour, the captain "unwittingly anchored in the foggy darkness, close to the lee shore of Stratton's Island." Without warning, "during the night, one of the most violent gales of the decade arose from the southwest"—tearing the vessel loose from both anchors and driving her hard against the island's jagged shore. Eyewitnesses stated that when the fog lifted on the morning of June 12, the vessel was a helpless shambles—her masts torn away. Only the hull was faintly visible from a distance of two miles at the nearest observation point on the cliffs at Prout's Neck. On hearing of the tragedy, Homer rushed out of his studio grabbed a handy academy board and made a simple, hurried sketch. Even with his still keen eyesight he could barely discern the outline of the hull. Later, as the debris washed up on Old Orchard Beach he collected a few spars and the ship's helm, placing them in his garden at Eastern Point. The wheel went over his mantel in the studio. For years the rough pencil or crayon sketch, regarded by the artist as a mere note, was set aside. Then, finally, Homer employing ". . . freely his gift for dramatization of a concept to the benefit of the picture's power as a work of art instead of a mere record" proceeded with the watercolor. "Formal and dramatic elements were combined by him" as he thought most advisable "to give his composition the fullest possible clarity and vividness of effect." Beam feels it is "fitting that this watercolor of a subject so close to his interests emerged as one of his strongest."

The finished watercolor, though not signed or dated, has been attested by members of Homer's family to be the last watercolor he ever completed, and if so, would have been finished about 1908. It expresses the death of the crew. Perhaps, he felt that death was waiting for him. It was one of the last important pictures he rendered in a medium which, more than any other American artist, he had raised to a level of major stature, and undoubtably was one of his finest paintings.

Bibliography

BOOKS

Barbour, G. M. *Florida for Tourists, Invalids & Settlers.* Gainsville: University of Florida Press, facsimile reproduction, 1964.

Barker, Virgil. *American Painting.* New York: Macmillan Co., 1950.

Baur, John I. H. *American Painting in the 19th Century.* New York: Praeger Publishers, 1953.

Beam, Philip C. *The Language of Art.* New York: Ronald Press, 1958.

Boswell, Peyton, Jr. *Modern American Painting.* New York: Dodd, Mead & Co., 1940.

Brooks, Edith G. *Saga of Baron Frederick DeBary & DeBary Hall, Florida.* Florida: Convention Press, Inc., 1966.

Caffin, Charles H. *Story of American Painting.* New York: Frederick A. Stokes Co., 1907.

Eliot, Alexander. *Three Hundred Years of American Painting.* New York: Time, Inc., 1957.

Hillyard, M. B. *The New South, A Book on the Resources, Attractions, and Development of the Southern States.* Baltimore: Record Printing House, 1885.

Isham, Samuel. *History of American Painting.* New York: Macmillan Co., 1936.

Gardner, Albert Ten Eyck. *History of Water Color Painting in America.* New York: Reinhold Publishing Co., 1966.

Munroe, C. K. *The Florida Annual Impartial and Unsectional, 1884.* New York: 1883.

Richardson, E. P. *Painting in America from 1502 to the Present.* New York: Thomas Y. Crowell Co., 1965.

Vaughn, Malcolm; Canaday, John; and Cheney, Sheldon. *Great Painters and Great Paintings.* New York: Readers' Digest Association, Inc., 1965.

Wallace, Willard M. *Soul of the Lion.* New York: Thomas Nelson, 1960.

MONOGRAPHS

Beam, Philip C. *Winslow Homer at Prout's Neck.* Boston: Little, Brown and Co., 1966.

Downes, William Howe. *The Life and Works of Winslow Homer.* New York and Boston: Houghton Mifflin Co., 1911.

Flexner, James Thomas. *The World of Winslow Homer 1836-1910.* New York: Time, Inc., 1966.

Gardner, Albert Ten Eyck. *Winslow Homer American Artist: His World and His Work.* New York: Clarkson N. Potter, Bramhall House, 1961.

Gelman, Barbara. *The Wood Engravings of Winslow Homer.* New York: Crown Publishers, Bounty Books, 1969.

Goodrich, Lloyd. *Winslow Homer.* New York: George Braziller, Inc., 1959.

______. *The Graphic Art of Winslow Homer.* Milan, Italy: Chanticleer Press, Inc., 1968.

Gould, Jean. *Winslow Homer, A Portrait.* New York: Dodd, Mead & Co., 1962.

Hoopes, Donelson F. *Winslow Homer Watercolors.* New York: Watson-Guptill Publications, 1969.

Watson, Forbes. *Winslow Homer.* New York: Crown Publishers, 1942.

Wilmerding, John. *Winslow Homer.* New York: Praeger Publishers, 1972.

PERIODICALS

Brinton, Christian. "Winslow Homer." *Scribner's Magazine,* January, 1911.

Buranelli, Marguerite. "Boy Wanted: Winslow Homer 1836-1910." *Highlights for Children,* 1969.

Chase, J. Eastman. "Some Recollections of Winslow Homer." *Harper's Weekly,* October 22, 1910.

Church, William C. "A Midwinter Resort, with Engravings of Winslow Homer's Watercolor Studies in Nassau." *The Century Magazine,* February, 1887.

Cox, Kenyon. "The Art of Winslow Homer." *Scribner's Magazine,* September, 1914.

Garrels, Harriet. *"Key West, Cabins & Palm Trees:* Painted by Winslow Homer." *The Instructor,* June, 1958.

Goodrich, Lloyd. "Pictorial Poet of the Sea and Forest." *Perspectives U.S.A.,* no. 14 (1956).

______. "Winslow Homer." *The Arts,* October 6, 1924.

Graham, Lois Homer. "An Intimate Glimpse of Winslow Homer's Art." *Vassar Journal of Undergraduate Studies,* May, 1936.

Hoeber, Arthur. "Winslow Homer, A Painter of the Sea." *The World's Work,* February, 1911.

Jewell, Edward Alden. "Winslow Homer Centenary Exhibition." *The New York Times,* December 20, 1936.

Katz, Leslie. "The Modernity of Winslow Homer." *Arts,* February, 1959.

Mather, Frank Jewett, Jr. "The Art of Winslow Homer." *Nation,* March 2, 1911.

McCausland, Elizabeth. "Winslow Homer—Graphic Artist." *Print,* April, 1937.

Mechlin, Leila. "Winslow Homer." *International Studio,* June, 1908.

Porter, Fairfield. "Homer—American vs. Artist: A Problem in Identities." *Art News,* December, 1958.

Rhinelander, David. "Winslow Homer Loved Cape Ann." *Gloucester Daily Times,* December 12, 1962.

Sleeper, Frank; and Cantwell, Robert. "Angler." *Sports Illustrated,* December 25, 1972.

Smith, Jacob Getlar. "The Watercolors of Winslow Homer." *American Artist,* February, 1955.

Vaughan, Malcolm. "He Spurned Success—and Achieved Fame." *The Reader's Digest,* July, 1955.

Watson, Forbes. "Winslow Homer." *American Magazine of Art,* October, 1936.

Weitenkampf, Frank. "The Intimate Homer." *Art Quarterly,* no. 6 (1943).

EXHIBITION CATALOGS

Albright-Knox Art Gallery. New York. "Watercolors by Winslow Homer." Gordon M. Smith. July 7-Aug. 28, 1966.

Babcock Galleries. New York. "Exhibition of Paintings by Winslow Homer." Carmine Dalesio. Dec. 8-31, 1941.

Brooklyn Museum. New York. "Watercolors by Winslow Homer." Oct. 16-Nov. 7, 1915.

Bowdoin College Museum of Art. Maine. "Winslow Homer at Prout's Neck." Philip C. Beam. 1966.

Buffalo Fine Arts Academy/Albright Art Gallery. New York. "An Important Group of Paintings in Oil and Watercolor by Winslow Homer." Dec. 15, 1929-Jan. 6, 1930.

Carnegie Institute. Pennsylvania. "Winslow Homer." Royal Cortissoz. Sept. 9—Oct. 26, 1923.

Castano Galleries. Massachusetts. "Watercolors and Etchings by Winslow Homer." Nov. 20-Dec. 16.

Colby College Press. Maine. "The Harold Trowbridge Pulsifer Collection of Winslow Homer Paintings and Drawings at Colby College." Samuel M. Green. 1949.

Colby and Bowdoin Colleges. Maine. "The Art of Winslow Homer." Nov. 1-21, 1954 and Dec. 1-21, 1954.

Cummer Gallery. Florida. "______." J. J. Dodge.

Doll & Richards Gallery. Massachusetts. "Watercolors by Winslow Homer." March 17-29, 1899.

Fogg Art Museum. Massachusetts. "Watercolors by Winslow Homer."

Friends of the Boston Symphony Orchestra. Massachusetts. "Exhibition of Paintings by Winslow Homer." George Henry Lovett Smith. Dec. 1, 1937.

International Arts Guild of the Bahamas. Nassau. "Winslow Homer (1836-1910)." Winslow Ames. March 2-24, 1962.

Lowe Art Museum, University of Miami. Florida. "Winslow Homer's Sub-tropical America." August L. Freundlich. 1968.

Macbeth Galleries. New York. "An Introduction to Homer." Dec. 15, 1936-Jan. 18, 1937.

Maynard Walker Gallery. New York. "Early Winslow Homer—Covering Twenty-five Years from 1864 to 1889." Exhibition Oct. 19-Nov. 6, 1953.

Mount Holyoke Friends of Art/Dwight Art Memorial. "Water Colors by Winslow Homer." Robert W. Macbeth. May 11-June 10, 1940.

Museum of Fine Arts. Massachusetts. "Winslow Homer, a Retrospective Exhibition." 1959.

Pennsylvania Academy of Fine Arts. Pennsylvania. "Sixth Annual Water Color Exhibit." 1908.

Prout's Neck Association. Maine. "Century Loan Exhibition As A Memorial to Winslow Homer, 1936." Booth Tarkington and Robert W. Macbeth.

Rhode Island School of Design. Rhode Island. "Exhibition of Water Colors by Winslow Homer." Feb. 6-March 1, 1931.

Smith College Museum of Art. Massachusetts. "Winslow Homer Illustrator 1860-1875." Mary Bartless Cowdrey and H. R. Hitchcock. 1951.

University Gallery, University of Florida. Florida. "Artists of the Florida Tropics." Roy C. Craven, Jr. 1965.

Walker Art Center. Minnesota. "American Watercolor and Winslow Homer." Lloyd Goodrich. 1945.

Whitney Museum of American Art. New York. "Winslow Homer." Lloyd Goodrich. 1973.

Worcester Art Museum. Massachusetts. "Winslow Homer." Nov. 16-Dec. 17, 1944.

Index